Acknowledgements

I wish to express my deep gratitude to my mentor, Professor Ilpo Laine for his valuable advice and constant encouragement. His inspiring lectures draw my attention to Nevanlinna theory in the first place, and his experienced guidance had a great impact on my work.

I am grateful to Dr. Rod Halburd and Professor Shun Shimomura for reading the manuscript and making many valuable comments and suggestions. Professor Aimo Hinkkanen also kindly read the manuscript pointing out several inaccuracies.

I give my warm thanks to the staff of the Joensuu mathematics department for providing me with a friendly environment, and for all the help they offered me.

For the financial assistance I am indebted to the Graduate School on Mathematical Analysis and Logic.

Last, but surely not the least, I wish to thank my family and all my friends for their support and encouragement. Especially, I am thankful to Joanna for standing by my side in these crucial times.

Joensuu, September 2002 Risto Korhonen

Contents

Introduction

The theory of *differential equations* is an essential part of analysis and calculus, which for over 300 years have been two of the most important branches of mathematics. In addition, since differential equations can be used to create effective mathematical models for many real world systems, they have significant applications in natural sciences and technology. These facts made Sophus Lie, a nineteenth century Norwegian mathematician, describe the importance of differential equations as follows:

> *"Among all the disciplines of mathematics, the* theory of differential equations *is the most important one. All areas of physics pose problems which lead to the integration of differential equations. In fact, it is the theory of differential equations which shows the way to understanding all time-dependent natural phenomena. If, on the one hand, the theory of differential equations has extreme* practical *significance, then, on the other hand, it attains a corresponding* theoretical *importance because it leads in a rational way to the study of new functions or new classes of functions."*

The history of differential equations begins from Leibniz and Newton. When they had acquired sufficient understanding and notation for the concept of derivative, it soon began to appear in equations. However, the foundations of this subject lie elsewhere. Leonhard Euler's affect in the development of differential equations was maybe the most substantial one. Among other things, he introduced general procedures to solve many different kinds of equations, and, by using numerical methods, he developed a way to find approximate solutions for nearly all differential equations. His contributions took the theory of differential equations past its primitive beginnings, and gave the future mathematicians a solid theoretical ground for further research.

One problem, already encountered in the early days of differential equations, has persisted over the years. Namely, in many cases, it is impossible to find an explicit solution for a given differential equation; solving techniques usually work in some limited special cases, and numerical methods yield only approximations. Nevanlinna theory offers an efficient way to bypass this problem, since it allows the study of the properties of a solution without knowing its explicit form. The only requirement needed, in order to apply this method, is that the solution must be meromorphic, either in the whole complex plane, or in a smaller domain where the growth of the solution is sufficiently fast, near the boundary of the domain. H. Wittich, whose research in this area began in 1942, was the first to make systematic studies in the applications of Nevanlinna theory into complex differential equations. Later on, due to the efforts of several mathematicians, including S. Bank and I. Laine, Nevanlinna theory became the leading tool in analyzing solutions of complex differential equations.

Like differential equations, the theory of *difference equations* has a wide range of applications in both mathematics and physics. Applications can be found, for example, by studying discretization methods for differential equations. This, in many occasions, leads to results in the theory of difference equations, which are natural discrete analogues of corresponding results of differential equations. These results can be further applied especially when considering discontinuous processes which appear, for instance, in the theory of probability, in the study of electrical networks, or in statistical problems.

Although differential and difference equations have some analogous features, there are significant differences between them. Firstly, solutions of polynomial differential equations have, quite often, very complicated branching structures, while large classes of difference equations admit mermomorphic solutions in the whole complex plane. Secondly, there are solutions of difference equations which do not satisfy any differential equation. For example, the Gamma function Γ does not satisfy any differential equation over the field of rational functions, but the linear homogeneous difference equation

$$f(z+1) - zf(z) = 0$$

has Γ as a solution. Hence it is interesting to study difference equations also in purely mathematical sense.

In this thesis, we consider differential equations and difference-differential equations from the point of view of Nevanlinna theory. We start by recalling the basic notations of Nevanlinna theory, as well as some elementary results from real analysis and from complex differential equations in Section 1. The rest of the paper can be divided into three parts as follows:

Firstly, we study algebraic differential equations in Section 2. We obtain information about the deficient slowly moving targets of solutions of differential equations in the general case, and more specific results in some special cases, such as the homogeneous and the nonhomogeneous cases. Moreover, we find an interesting connection between deficient slowly moving targets of a meromorphic function and hypertranscendency of the function.

Secondly, we consider algebraic differential equations of the first and the second order in Sections 3 and 4, respectively. In the first order case, we focus on the behavior of the generalized Malmquist differential equations. In particular, we study the value distribution of solutions of Riccati differential equations with rational coefficients. In Section 4, we concentrate mainly on the value distribution of the Painlevé transcendents, but also second order linear homogeneous differential equations are briefly examined.

Finally, in Section 5, we apply Nevanlinna theory to obtain some information about solutions of algebraic difference-differential equations. Among other things, we obtain a theorem which can be described as an analogue of the lemma of the logarithmic derivative for difference-differential equations. This theorem is further applied to prove a lemma of Clunie type for these equations.

1. Notations and background

In this section, we recall some definitions and preliminary results from Nevanlinna theory, real analysis, and from the theory of complex differential equations. We note that by a meromorphic function we always mean a function which is meromorphic in the whole complex plane.

1.1. Nevanlinna theory

We use the standard notations of Nevanlinna theory, i.e. $m(r,f)$, $n(r,f)$, $N(r,f)$ and $T(r,f)$ denote the *proximity function*, the *non-integrated counting function*, the *counting function* and the *characteristic function* of f, respectively. Furthermore, the *order of growth* and the *lower order* of a meromorphic function f are denoted by $\rho(f)$ and $\mu(f)$, respectively. By $\lambda(f)$ we mean the *exponent of convergence* of the zero sequence of f, and similarly $\lambda(1/f)$ denotes the exponent of convergence of the pole sequence of f. Moreover, we say that meromorphic function g is *small* with respect to f if $T(r,g) = S(r,f)$, where $S(r,f) = o(T(r,f))$ outside of a possible exceptional set of finite linear measure. We also call this kind of function a *slowly moving target* with respect to f. If there is no exceptional set, we use the notation $o(T(r,f))$ to make a distinction with $S(r,f)$.

The *deficiency* of f with respect to g is defined by

$$\delta(g,f) := \liminf_{r\to\infty} \frac{m\left(r, \frac{1}{f-g}\right)}{T(r,f)},$$

and g is called a *deficient function* with respect to f if $\delta(g,f) > 0$. Similarly, the *Valiron deficiency* of f with respect to g is defined by

$$\Delta(g,f) := \limsup_{r\to\infty} \frac{m\left(r, \frac{1}{f-g}\right)}{T(r,f)},$$

and g is called a *Valiron deficient function* if $\Delta(g,f) > 0$.

Denote

$$\overline{N}\left(r,\frac{1}{f-a}\right) := \int_0^r \frac{\overline{n}(t,a)-\overline{n}(0,a)}{t}dt + \overline{n}(0,a)\log r,$$

where every a-point in $\overline{n}(t,a)$ is counted only once, independently of its multiplicity. Then the *total ramification index* of f with respect to g is defined by

$$\Theta(g,f) := 1 - \limsup_{r\to\infty} \frac{\overline{N}\left(r,\frac{1}{f-g}\right)}{T(r,f)}.$$

Furthermore, in order to obtain counting functions for multiple points, denote

$$\begin{aligned}
N_1\left(r,\frac{1}{f-a}\right) &:= N\left(r,\frac{1}{f-a}\right) - \overline{N}\left(r,\frac{1}{f-a}\right),\\
N_1(r,f) &:= 2N(r,f) - N(r,f'),\\
N_B(r,f) &:= N\left(r,\frac{1}{f'}\right) + 2N(r,f) - N(r,f'),\\
\vartheta(g,f) &:= \liminf_{r\to\infty} \frac{N_1\left(r,\frac{1}{f-g}\right)}{T(r,f)},\\
\Phi(f) &:= \liminf_{r\to\infty} \frac{N_B(r,f)}{T(r,f)},\\
\Phi_e(f) &:= \liminf_{r\to\infty} \frac{N\left(r,\frac{1}{f'}\right)}{T(r,f)}.
\end{aligned}$$

Finally, a transcendental meromorphic function f is called *admissible* with respect to $a_j(z)$, $j=1,\dots,n$, if $T(r,a_j) = S(r,f)$ for all $j=1,\dots,n$.

The following theorem is an immediate consequence of the Poisson-Jensen formula. The proof can be found, for example, in [21, p. 26].

Theorem 1.1.1. *Let f be an entire function, and assume that $0<r<R<\infty$ and that the maximum modulus $M(r,f) := \max_{|z|=r}|f(z)|$ satisfies $M(r,f)\geq 1$. Then*

$$T(r,f) \leq \log M(r,f) \leq \frac{R+r}{R-r}T(R,f).$$

In order to deal with logarithmic derivatives, we need two theorems. But first, define the *central index* $\nu(r,f)$ of an entire function

$$f(z) = \sum_{n=0}^{\infty} a_n z^n$$

as the greatest exponent m such that

$$|a_m|r^m = \max_{n\geq 0} |a_n|r^n.$$

The proof of the following theorem can be found, for example, in [20, pp. 197–199].

Theorem 1.1.2 (Wiman-Valiron). *Let f be transcendental entire function, let $0 < \delta < \frac{1}{4}$, and let z, $|z| = r$, be such that*

$$|f(z)| > M(r,f)\nu(r,f)^{-\frac{1}{4}+\delta}$$

holds. Then there exists a set $F \subset \mathbb{R}_+$ of finite logarithmic measure, i.e. $\int_F dt/t < +\infty$, such that

$$f^{(m)}(z) = \left(\frac{\nu(r,f)}{z}\right)^m (1+o(1))f(z)$$

holds for all $m \geq 0$ and $r \notin F$.

The next theorem is due to Gundersen, see [10].

Theorem 1.1.3 (Gundersen). *Let k, j be integers, such that $k > j \geq 0$, let f be a meromorphic function of finite order $\rho(f)$, and let $\varepsilon > 0$ be a given constant. Then there exists a set $E \subset (1,\infty)$ that has a finite logarithmic measure, such that for all z satisfying $|z| \notin E \cup [0,1]$, we have*

$$\left|\frac{f^{(k)}}{f^{(j)}}\right| \leq |z|^{(k-j)(\rho(f)-1+\varepsilon)}.$$

We also need a classical theorem due to Hadamard, which is very useful in considering meromorphic functions of finite order.

Theorem 1.1.4 (Hadamard). *Let f be a non-constant entire function of finite order ρ. Then*

$$f(z) = z^m Q(z) e^{P(z)},$$

where

1. *$m \geq 0$ is the multiplicity of the zero of f at $z = 0$,*

2. *$Q(z)$ is the canonical product formed with nonzero zeros of f,*

3. *$P(z)$ is a polynomial of degree $\leq \rho$.*

We need another lemma concerning the exponent of convergence and the order of growth of a canonical product.

Lemma 1.1.5. *For a canonical product* $Q(z)$, $\lambda(Q) = \rho(Q)$.

To consider the exponent of convergence for solutions of some difference equations, the following simple lemma appears to be useful.

Lemma 1.1.6. *Let* f *be a meromorphic function. Then*

$$\lambda(f(z)) = \lambda(f(z+c)),$$

and

$$\lambda\left(\frac{1}{f(z)}\right) = \lambda\left(\frac{1}{f(z+c)}\right),$$

where $c \in \mathbb{C}$.

Proof. Let $(b_n)_{n\in\mathbb{N}}$ be the zero sequence of $f(z)$. Then $(b_n - c)_{n\in\mathbb{N}}$ is the zero sequence of $f(z+c)$. Since for any $\nu > 0$,

$$\lim_{n\to\infty}\left(\frac{|b_n - c|^{-\nu}}{|b_n|^{-\nu}}\right) = 1,$$

the sum $\sum_{n=1}^{\infty} |b_n - c|^{-\nu}$ converges if and only if $\sum_{n=1}^{\infty} |b_n|^{-\nu}$ converges, by the standard limit comparison test of real analysis. Hence, $\lambda(f(z)) = \lambda(f(z+c))$. A similar reasoning applies for the poles of f. □

If for a meromorphic function f,

$$\rho(f) = \mu(f),$$

then f is said to be of *regular growth*.

We close this section with a lemma concerning the growth of meromorphic functions with Borel exceptional zeros and poles. Even though the proof is simple, the lemma itself plays an important role in considering meromorphic functions of finite order with few zeros and poles in the Borel sense.

Lemma 1.1.7. *Let* f *be a meromorphic function of order* ρ. *If* f *satisfies*

$$\max\left(\lambda(f), \lambda\left(\frac{1}{f}\right)\right) < \rho, \tag{1.1}$$

then f *is of regular growth.*

Proof. Since the zeros and poles of f are Borel exceptional by (1.1), we may apply a result due to Whittaker, see [20, Satz 13.4], to deduce that f is of regular growth. □

1.2. Real analysis

We start with a lemma which allows us to deal with exceptional sets in some situations. Even though the lemma itself considers the Nevanlinna characteristic function, it is nothing more than a modification of a lemma from real analysis. Since the proof is slightly modified to what can be found, say, in [21, p. 5], we include the details here for convenience.

Lemma 1.2.1. *Let g and h be meromorphic functions such that*

$$T(r,g) = S(r,h). \tag{1.2}$$

Then, for any $\alpha > 1$, there exists $r_0 > 0$ such that $T(r,g) = o(T(\alpha r, h))$ for all $r > r_0$.

Proof. Denote $\sigma := \int_E dr$, where E is the exceptional set associated with equation (1.2), in other words, $T(r,g)$ is of growth $o(T(r,h))$ everywhere outside E. Choose $r_0 := \sigma/(\alpha - 1)$. Then since for any $r > r_0$,

$$\int_r^{\alpha r} dt = r(\alpha - 1) > r_0(\alpha - 1) = \sigma,$$

the interval $[r, \alpha r]$ meets the complement of E for all $r > r_0$. Hence, taking $t \in [r, \alpha r] \backslash E$, we get

$$T(r,g) \leq T(t,g) \leq \varepsilon T(t,h) \leq \varepsilon T(\alpha r, h)$$

for all $\varepsilon > 0$, when r is large enough. Hence,

$$\frac{T(r,g)}{T(\alpha r, h)} \leq \varepsilon$$

for all $\varepsilon > 0$, when r is large enough, and so the assertion follows. □

The following lemma is needed to get rid of an exceptional set of finite logarithmic measure. The proof can be found in [8].

Lemma 1.2.2. *Let g and h be monotone non-decreasing functions on $[0, \infty)$ such that $g(r) \leq h(r)$ for all $r \notin E$ where $E \subset (1, \infty)$ is a set of finite logarithmic measure. Let $\alpha > 1$ be a given constant. Then there exists an $r_0 = r_0(\alpha) > 0$ such that $g(r) \leq h(\alpha r)$ for all $r \geq r_0$.*

1.3. Complex differential equations

We denote by $\mathcal{M}$ the field of meromorphic functions, and by $\mathcal{L}$ a subfield of $\mathcal{M}$. Let f be a given non-constant meromorphic function, and let

$$\mathcal{L}_f := \{h \text{ meromorphic} : T(r,h) = S(r,f)\}.$$

The set $\mathcal{L}_f$ equipped with the operations of pointwise multiplication, addition and complex derivation forms a *differential field.* Moreover, we denote

$$\begin{aligned} \mathcal{L}_0 &:= \{h \text{ meromorphic} : T(r,h) = o(r)\}, \\ \mathcal{L}_{f,0} &:= \{h \text{ meromorphic} : T(r,h) = o(T(r,f))\}, \\ \mathcal{L}_\rho &:= \{h \text{ meromorphic} : \rho(h) < \rho\}. \end{aligned}$$

Note that functions in the field $\mathcal{L}_{f,0}$ are small with respect to f without exceptional set. Hence, $\mathcal{L}_{f,0} \subset \mathcal{L}_f$. Finally, by $\mathbb{C}(z)$ we mean the differential field of rational functions.

We say that a meromorphic function f is *differentially algebraic* with respect to a differential field $\mathcal{L}$ if it satisfies an algebraic differential equation

$$P\left(z, w, w', \ldots, w^{(n)}\right) = 0 \tag{1.3}$$

over $\mathcal{L}$. We denote the family of all such functions by $\mathcal{A}(\mathcal{L})$. Note that for convenience of notation, we often use $P(z,w)$, instead of the left hand side of (1.3). If f does not satisfy any equation of the type (1.3), it is *hypertranscendental* with respect to $\mathcal{L}$. We can write (1.3) in the form

$$\sum_{\lambda=(j_0,\ldots,j_n)\in I} P_\lambda(z) w^{j_0}(w')^{j_1}\cdots(w^{(n)})^{j_n} = 0, \tag{1.4}$$

where I is a finite set of *multi-indices* λ, and $P_\lambda \in \mathcal{L}$ for all $\lambda \in I$. Then the *degree* of a single term in (1.4) is defined by

$$d(\lambda) := j_0 + j_1 + \cdots + j_n,$$

and its *weight* by

$$w(\lambda) := j_1 + 2j_2 + \cdots + nj_n.$$

Let $\alpha = (a_0, \ldots, a_n)$ and $\beta = (b_0, \ldots, b_n)$ be two multi-indices. We say that β is *higher* than α, or $\alpha < \beta$, if any one of the following conditions is satisfied:

(i) $d(\alpha) < d(\beta)$,

(ii) $d(\alpha) = d(\beta)$, but $w(\alpha) < w(\beta)$,

(iii$_n$) $d(\alpha) = d(\beta)$ and $w(\alpha) = w(\beta)$, but $a_n < b_n$, (iii$_{n-1}$) $d(\alpha) = d(\beta)$, $w(\alpha) = w(\beta)$ and $a_n = b_n$, but $a_{n-1} < b_{n-1}$,

$\vdots$

(iii$_0$) $d(\alpha) = d(\beta)$, $w(\alpha) = w(\beta)$, $a_n = b_n$, ..., $a_1 = b_1$, but $a_0 < b_0$.

Remark. In fact, the conditions (iii$_0$) and (iii$_1$) above are not needed, since, in (iii$_0$), $d(\alpha) = d(\beta)$ yields $a_0 = b_0$; and in (iii$_1$), $d(\alpha) = d(\beta)$ and $w(\alpha) = w(\beta)$ yield $a_0 = b_0$ and $a_1 = b_1$.

If none of the conditions above is satisfied and $\alpha \neq \beta$, then clearly $\alpha > \beta$. The *principal term* of a differential polynomial P is the term of highest index. Given two differential polynomials P and Q, we say that P is *higher* than Q, or $P > Q$, if the index of the principal term of P is higher than the index of the principal term of Q. Moreover, we define the degree, $d(P)$, and the weight, $w(P)$, of P as the degree and the weight of the principal term of P, respectively.

Let f be a differentially algebraic function. Then, out of all differential polynomials $P_\nu(z, w)$, such that $P_\nu(z, f) = 0$, let $M(z, w)$ be minimal with respect to the ordering described above. We may assume that the coefficient of the principal term is one. Then this polynomial is unique, and we call it the *minimal polynomial* of f.

We need the following theorem in order to handle some differentially algebraic functions. The proof of this theorem can be found, for example, in [21, pp. 289–290].

Theorem 1.3.1. *Suppose that* $y \in \mathcal{A}(\mathcal{A}(\mathcal{L}))$. *Then* $y \in \mathcal{A}(\mathcal{L})$.

We also need the following application of the lemma of the logarithmic derivative. It is a special case of [22, Theorem 6] (See also [21, Proposition 9.2.3]).

Lemma 1.3.2 (Mohon'ko). *Let* $P(z, u)$ *be a polynomial in* u *and its derivatives with meromorphic coefficients* $\{b_\nu : \nu \in N\}$. *Assume that* $u = f$ *is a non-constant meromorphic solution of the differential equation* $P(z, u) = 0$, *and that* $T(r, b_\nu) = S(r, f)$ *for all* $\nu \in N$. *If* $P(z, 0) \not\equiv 0$, *then* $m(r, 1/f) = S(r, f)$.

The following result is a very useful tool while considering complex differential equations. It is a version of a theorem due to S. Hellerstein and L. A. Rubel, and the proof follows by a careful examination of the proof of [14, Theorem 1].

Theorem 1.3.3 (Hellerstein and Rubel). *Let* f *and* g *be meromorphic functions. If*

$$a_n(z)g^n + \cdots + a_1(z)g + a_0(z) = 0,$$

where $a_j \in \mathcal{L}_f$ *for all* $j = 0, \ldots, n$, *then* $g \in \mathcal{L}_f$.

Theorem 1.3.3 says that the field $\mathcal{L}_f$ is algebraically closed in $\mathcal{M}$, in other words, any meromorphic function satisfying an algebraic equation over the field $\mathcal{L}_f$ belongs to $\mathcal{L}_f$.

2. Algebraic differential equations of arbitrary order

The value distribution of solutions of algebraic differential equations is an area investigated rather actively over the last three decades by many authors. Nevertheless, a lot remains to be done in this large, yet interesting, subfield of complex function theory.

We start our considerations by taking a look at the general case, and, later on, we restrict our studies to some special cases, like homogeneous and nonhomogeneous algebraic differential equations, and algebraic differential equations with finite order solutions. Finally, we say a few words about hypertranscendental functions.

2.1. Deficient slowly moving targets

The deficient values of a differentially algebraic meromorphic function f are strongly linked to the algebraic differential equation $P(z,u)=0$ having f as a solution. A. Z. Mohon'ko and V. D. Mohon'ko proved that if $P(z,c)\not\equiv 0$ for a constant $c\in\mathbb{C}$, then $\delta(c,f)=0$ [22]. The following theorem is a generalization of this result for slowly moving targets.

Theorem 2.1.1. *Let f be a transcendental meromorphic function, which is differentially algebraic with respect to $\mathcal{L}_f$ with the minimal polynomial $M(z,w)$. Then*

$$\delta(g,f)=0, \tag{2.1}$$

whenever $g(z)$ is a small function with respect to f in the sense $T(r,g)=S(r,f)$, such that $w=g$ does not satisfy

$$M(z,g,g',\ldots,g^{(n)})=0.$$

Proof. By substituting $w=W+g$ into the equation

$$M(z,w,w',\ldots,w^{(n)})=0,$$

we have an equation of the form

$$M(z,W+g):=M(z,W+g,W'+g',\ldots,W^{(n)}+g^{(n)})=0.$$

Write

$$M(z,W+g)=\sum_{\lambda=(j_0,\ldots,j_n)\in I} P_\lambda(z)(W+g)^{j_0}(W'+g')^{j_1}\cdots(W^{(n)}+g^{(n)})^{j_n},$$

and divide this sum into two parts as follows:

$$\begin{aligned} M(z, W+g) = & \sum_{\substack{\lambda=(j_0,\dots,j_n)\in I' \\ d(\lambda)\geq 1}} Q_\lambda(z) W^{j_0}(W')^{j_1}\cdots(W^{(n)})^{j_n} \\ & + \sum_{\lambda=(j_0,\dots,j_n)\in I} P_\lambda(z) g^{j_0}(g')^{j_1}\cdots(g^{(n)})^{j_n} \\ = & \sum_{\substack{\lambda=(j_0,\dots,j_n)\in I' \\ d(\lambda)\geq 1}} Q_\lambda(z) W^{j_0}(W')^{j_1}\cdots(W^{(n)})^{j_n} \\ & + M(z, g, g', \dots, g^{(n)}) = 0, \end{aligned} \tag{2.2}$$

where the coefficients Q_λ are polynomials in $P_\lambda, g, g', \dots, g^{(n)}$ for all $\lambda \in I' \supset I$, and so they are also small with respect to f. Now $W = f - g$ is a solution of (2.2). Since by assumption,

$$M(z, g, g', \dots, g^{(n)}) \not\equiv 0,$$

we deduce by Lemma 1.3.2 that

$$m\left(r, \frac{1}{W}\right) = m\left(r, \frac{1}{f-g}\right) = S(r, f-g).$$

Hence $m(r, 1/W) = S(r, f)$, and so $\delta(g, f) = 0$. □

Remark. By a careful inspection of the above proof, it is easily seen that the condition $M(z, g) \not\equiv 0$ can be replaced by $P(z, g) \not\equiv 0$, where P is any differential polynomial such that $P(z, f) \equiv 0$.

The following example demonstrates the use of Theorem 2.1.1.

Example 2.1.2. The function $\psi(z) := (\sin z)^{-1}$ is differentially algebraic with respect to $\mathbb{C}(z)$, since it is a solution of the algebraic differential equation

$$ff'' - f^2 - 2f'^2 = 0. \tag{2.3}$$

We start by showing that (2.3) is the minimal polynomial of ψ over $\mathbb{C}(z)$.

Denote the minimal polynomial of ψ by $M(z, f)$. Suppose first that $M(z, f)$ is of degree one. Then $M(z, f)$ is of the form

$$M(z, f) = f^{(n)} + a_{n-1}(z) f^{(n-1)} + \cdots + a_0(z) f + R(z) = 0, \tag{2.4}$$

where $a_j(z)$, $j = 0, 1, \dots, n-1$, and $R(z)$ are rational functions. Since ψ has infinitely many poles, we can choose a pole z_0 of ψ which is not a pole of any of the coefficient functions. But now, since all poles of ψ are simple, we have a contradiction by comparing the orders of the poles of the terms of (2.4) at z_0. Hence the minimal polynomial of ψ must be at least of degree two. From the equation (2.3) it follows that $M(z, f)$ is exactly of degree two.

Suppose now that $M(z, f)$ has weight one. Then, since the degree of $M(z, f)$ is two,

$$M(z, f) = ff' + a_3(z)f' + a_2(z)f^2 + a_1(z)f + a_0(z) = 0,$$

where a_j, $j = 0, \ldots, 3$, are rational functions. Similarly as to above, we again have a contradiction by comparing the orders of the poles at z_0. Hence also the weight of $M(z, f)$ is two.

We have narrowed all possible cases down to two: the principal term of $M(z, f)$ can be either $(f')^2$, or ff''. Suppose first that it is $(f')^2$. Then

$$M(z, f) = (f')^2 + a_4(z)ff' + a_3(z)f' + a_2(z)f^2 + a_1(z)f + a_0(z) = 0,$$

which again leads to a contradiction similarly as above. Hence the principal term of $M(z, f)$ is ff'', and so we have proved our claim, since the principal term determines the minimal polynomial uniquely.

Assume now that there exists a nonzero rational function g, which is a solution of $M(z, f) = 0$. Then g is of the form

$$g(z) = \frac{p(z)}{q(z)} = \frac{b_n z^n + b_{n-1}z^{n-1} + \cdots + b_0}{c_m z^m + c_{m-1}z^{m-1} + \cdots + c_0}, \tag{2.5}$$

where p and $q \not\equiv 0$ have no common factors, and $b_n, c_m \neq 0$. Here the coefficients $b_j, c_j \in \mathbb{C}$ are normalized by $c_m = 1$. Then, by substituting $g = p/q$ into (2.3), we have

$$p''pq^2 - q''qp^2 - p^2q^2 - 2(p')^2q^2 + 2pp'qq' \equiv 0. \tag{2.6}$$

Furthermore, by substituting (2.5) into (2.6), we can see that the leading coefficient of (2.6) is b_n^2. Hence $b_n = 0$, since the polynomial (2.6) is identically zero. But this contradicts our assumption, and therefore the only rational solution of $M(z, f) = 0$ is the zero solution. Thus by Theorem 2.1.1, $\delta(g, \psi) = 0$ for all rational functions $g \not\equiv 0$. Finally, since

$$N\left(r, \frac{1}{\psi}\right) = N(r, \sin z) = 0,$$

we have $\delta(0, \psi) = 1$.

2.2. Deficient slowly moving targets in the finite order case

Since many classical differential equations, like the first, the second and the fourth Painlevé equations, and the Riccati differential equation with polynomial coefficients, have meromorphic solutions of finite order, it is useful to study the properties of differentially algebraic functions of finite order. We start these considerations by a lemma of Clunie type. The proof can be found, for instance, in [11, pp. 68–69].

Lemma 2.2.1 (Clunie). *Let f be a transcendental meromorphic solution of finite order of the differential equation*

$$f^n P(z,f) = Q(z,f),$$

where $P(z,f)$ and $Q(z,f)$ are polynomials in f and its derivatives with rational coefficients. If the total degree of $Q(z,f)$ as a polynomial in f and its derivatives is $\leq n$, then

$$m\left(r, P(z,f)\right) = O(\log r).$$

Similarly, we state the following finite order version of Mohon'ko's lemma [22, Theorem 6].

Lemma 2.2.2 (Mohon'ko). *Let $P(z,u)$ be a polynomial in u and its derivatives with rational coefficients, and assume that $u = f$ is a transcendental meromorphic solution of finite order of the differential equation $P(z,u) = 0$. If $P(z,0) \not\equiv 0$, then $m(r, 1/f) = O(\log r)$.*

Remark. If we suppose that all assumptions of Lemma 2.2.2 hold, except that the coefficients of P belong to the field $\{h : T(r,h) = O(r)\}$, instead of being rational, and that $T(r,f) = O(e^{\Gamma r})$ for some $\Gamma > 0$, instead of f being of finite order, we have that $m(r, 1/f) = O(r)$.

We are now able to obtain a somewhat stronger version of Theorem 2.1.1 in the important special case where f is of finite order.

Theorem 2.2.3. *Let f be transcendental meromorphic function of finite order. If f is differentially algebraic with respect to $\mathbb{C}(z)$ with the minimal polynomial $M(z,w)$, then*

$$\Delta(g,f) = 0, \tag{2.7}$$

whenever $g(z)$ is a rational function not satisfying

$$M(z, g, g', \ldots, g^{(n)}) = 0. \tag{2.8}$$

Proof. By using Lemma 2.2.2 instead of Lemma 1.3.2 and slightly modifying the proof of Theorem 2.1.1 in relevant parts, we deduce that

$$m\left(r, \frac{1}{f-g}\right) = O(\log r)$$

for all rational g not satisfying (2.8). Hence,

$$\Delta(g, f) = \limsup_{r\to\infty} \frac{m\left(r, \frac{1}{f-g}\right)}{T(r, f)} = \limsup_{r\to\infty} \frac{O(\log r)}{T(r, f)} = 0,$$

since f was assumed to be transcendental. □

2.3. Nonhomogeneous differential equations

In this section, we study the value distribution of solutions of nonhomogeneous algebraic differential equations. We start with some terminology.

Let

$$P(z, u) = \sum_{\lambda=(j_0,\ldots,j_n)\in I} P_\lambda(z) u^{j_0} (u')^{j_1} \cdots (u^{(n)})^{j_n} \tag{2.9}$$

be a differential polynomial. We may write (2.9) in the form

$$P(z, u) = \sum_{j=0}^{m(P)} P_j(z, u),$$

where $m(P) \in \mathbb{N}$, and all the terms of P_j are of the same degree $d(P_j)$ for all $j = 0, \ldots, m(P)$. Then P_j is called as the *homogeneous component* of P of degree $d(P_j)$.

Definition 2.3.1. Let $P(z, u)$ be a differential polynomial. Then

1. $P(z, u)$ is a *homogeneous form* if there exists a meromorphic function $f \not\equiv 0$ such that, for every homogeneous component $P_j(z, u)$, the identity $P_j(z, f) \equiv 0$ holds.

2. $P(z, u)$ is *nonhomogeneous*, if it is not a homogeneous form, namely if there exists no meromorphic function $f \not\equiv 0$ satisfying $P_0(z, f) \equiv \cdots \equiv P_m(z, f) \equiv 0$.

Note that a homogeneous form is not always a homogeneous polynomial. For instance,

$$(u')^2 - u''u + u' - u$$

is a homogeneous form.

If we make an additional assumption restricting the number of poles and zeros of a differentially algebraic function, we obtain

Lemma 2.3.2. *Let f be a transcendental meromorphic function such that*

$$\max\left(\overline{N}(r,f),\overline{N}\left(r,\frac{1}{f}\right)\right)=S(r,f), \tag{2.10}$$

and let $P(z,u)$ be a differential polynomial with meromorphic coefficients $b_\nu \in \mathcal{L}_f$. If $P(z,f)\equiv 0$, then $P(z,u)$ is a homogeneous form, and for each homogeneous component, $P_j(z,f)\equiv 0$.

Proof. Assume conversely that there exists a nonhomogeneous differential polynomial P such that $P(z,f)\equiv 0$. Write P in the form

$$P(z,u)=\sum_{j=0}^{m}P_j(z,u),$$

where P_j is homogeneous component of P for each $j=0,\ldots,m$. Since we assumed that P is nonhomogeneous, $P_j(z,f)\not\equiv 0$ for some $j=0,\ldots,m$. Because $P(z,f)\equiv 0$, there must be at least two such P_j, and so $m>0$. Now $u=f$ also satisfies the differential equation

$$Q(z,u):=\sum_{j=0}^{m}\frac{P_j(z,f)}{f^{d(P_j)}}\cdot u^{d(P_j)}=0, \tag{2.11}$$

where $d(P_j)$ denotes the degree of P_j in u and its derivatives. Note that by (2.10),

$$N\left(r,\frac{f^{(n)}}{f}\right)\leq n\left(\overline{N}(r,f)+\overline{N}\left(r,\frac{1}{f}\right)\right)=S(r,f)$$

for all $n\in\mathbb{N}$. Thus,

$$N\left(r,\frac{P_j(z,f)}{f^{d(P_j)}}\right)=S(r,f),$$

and hence we can write (2.11) in the following form by using the lemma of the logarithmic derivative:

$$Q(z,u)=\sum_{j=0}^{l}a_j(z)u^j=0, \tag{2.12}$$

where $l\in\mathbb{N}$, $a_l(z)\not\equiv 0$, and $a_j\in\mathcal{L}_f$ for $j=0,\ldots,l$. By Theorem 1.3.3 $T(r,f)=S(r,f)$, since $Q(z,f)\equiv 0$. This is a contradiction which proves the assertion. □

The following theorem is an immediate consequence of Lemma 2.3.2.

Theorem 2.3.3. *Let f be a transcendental meromorphic function such that*

$$\Theta(0,f) = \Theta(\infty,f) = 1, \tag{2.13}$$

and let $P(z,u)$ be a nonhomogeneous polynomial in u and its derivatives with meromorphic coefficients $b_\nu \in \mathcal{L}_f$. Then $P(z,f) \not\equiv 0$.

Proof. By (2.13), we have

$$\max\left(\overline{N}(r,f), \overline{N}\left(r,\frac{1}{f}\right)\right) = o(T(r,f)),$$

and so the assertion follows by Lemma 2.3.2. □

The following theorem gives us a criterion when a nonhomogeneous differential polynomial can be replaced by a homogeneous form in a differential equation, after suitably transforming the solution of the original equation.

Theorem 2.3.4. *Let f be a meromorphic function, and let $P(z,u)$ be a differential polynomial with coefficients in the field $\mathcal{L}_f$, such that $P(z,f) \equiv 0$.*

1. *If there exist $g_j \in \mathcal{L}_f$, $j = 1,2$, such that*

$$\overline{N}\left(r,\frac{1}{f-g_j}\right) = S(r,f), \tag{2.14}$$

then there exists a Möbius transformation Φ with coefficients in $\mathcal{L}_f$, and a homogeneous form Q over $\mathcal{L}_f$, such that $Q(z,\Phi(f)) \equiv 0$.

2. *If $\overline{N}(r,f) = S(r,f)$, and there exists $g \in \mathcal{L}_f$, such that*

$$\overline{N}\left(r,\frac{1}{f-g}\right) = S(r,f),$$

then there exists a Möbius transformation Φ with coefficients in $\mathcal{L}_f$, and a homogeneous form Q over $\mathcal{L}_f$, such that $Q(z,\Phi(f)) \equiv 0$.

Proof. 1. If P is a homogeneous form, just take $Q \equiv P$ and $\Phi(f) \equiv f$. Hence we may assume that P is nonhomogeneous. By assumption there exist two functions $g_1, g_2 \in \mathcal{L}_f$ such that

$$\overline{N}\left(r,\frac{1}{f-g_j}\right) = S(r,f)$$

for $j = 1,2$. Then clearly the function

$$\Phi(f) := \frac{f-g_1}{f-g_2}$$

is differentially algebraic over $\mathcal{L}_f$ satisfying, say, an algebraic differential equation $Q(z, u) = 0$. Clearly also

$$\max\left(\overline{N}(r, \Phi(f)), \overline{N}\left(r, \frac{1}{\Phi(f)}\right)\right) = S(r, f) = S(r, \Phi(f)).$$

Hence, by Lemma 2.3.2, Q is a homogeneous form with homogeneous components such that $Q_j(z, \Phi(f)) \equiv 0$.

2. We still have to consider the case $\overline{N}(r, f) = S(r, f)$. By assumption there exist $g \in \mathcal{L}_f$ such that

$$\overline{N}\left(r, \frac{1}{f - g}\right) = S(r, f).$$

Denote $\Phi(f) := f - g$. The rest of the proof follows by completely similar reasoning as was done above. □

Previous investigations of nonhomogeneous differential equations, say e.g. of Painlevé differential equations, have shown that the value distribution of solutions of such equations tends to be of very "regular" nature. Theorem 2.3.4 demonstrates that if the value distribution of a differentially algebraic function f is, loosely speaking, "irregular", in the sense that (2.14) holds, then the differential equation, having f as a solution, is transformable into a homogeneous form by a suitable Möbius transformation.

In the following corollary, the criterion is given in terms of the total ramification index.

Corollary 2.3.5. *Let f be a meromorphic function satisfying an algebraic differential equation $P(z, u) = 0$. If there exist $g_1, g_2 \in \mathcal{L}_f$ such that*

$$\Theta(g_j, f) = 1 \tag{2.15}$$

for $j = 1, 2$, then there exists a homogeneous form $Q(z, u)$, such that $Q(z, \Phi(f)) \equiv 0$, where

$$\Phi(f) := \frac{f - g_1}{f - g_2}.$$

In fact, all algebraic differential equations having $\Phi(f)$ as a solution are homogeneous forms, in the sense that all of their homogeneous components must vanish identically when substituting $\Phi(f)$ into the equation.

Proof. If P is a homogeneous form there is nothing to prove. Hence, we may assume that P is nonhomogeneous. By assumption, $\Theta(g_j, f) = 1$ for $g_1, g_2 \in \mathcal{L}_f$. Thus, by the definition of Θ,

$$\overline{N}\left(r, \frac{1}{f - g_j}\right) = o(T(r, f))$$

for both $j = 1, 2$. Hence assertion follows by a careful inspection of the proof of Theorem 2.3.4. □

Remarks. 1. The assertion of Corollary 2.3.5 still holds if we replace the condition (2.15) by $\Theta(\infty, f) = \Theta(g_1, f) = 1$. The proof of this is just a minor adjustment of the above proof.

2. If we have a differential equation $P(z, u) = 0$ with a meromorphic solution f satisfying (2.14) or (2.15) for some g_1, g_2, then by substituting

$$u = \Phi^{-1}(f) = \frac{g_2(z)f - g_1(z)}{f - 1}$$

into the equation $P(z, u) = 0$, and by multiplying out denominators, we obtain an equation $F(z, u) = 0$, which has $\Phi(f)$ as a solution. Moreover, each homogeneous component of F must vanish identically when $\Phi(f)$ is substituted into it, and so Q in Theorem 2.3.4 and Corollary 2.3.5 can be chosen to be any of those components, provided that all coefficients of the component do not vanish identically after the substitution. Note also that if P is a homogeneous form to begin with, then Q, obtained by the transformation described above, has to be still a homogeneous form.

We will further demonstrate the use of Corollary 2.3.5 in Section 3.

2.4. Homogeneous differential equations

In the previous section we gave a condition under which a nonhomogeneous algebraic differential equation is transformable into a homogeneous form. Therefore, it is natural to study the behavior of solutions of homogeneous algebraic differential equations. In this section, we consider differential equations

$$P(z, f) := P(z, f, f', \ldots, f^{(n)}) = 0, \tag{2.16}$$

with polynomial coefficients. We may write (2.16) in the form

$$P(z, f) = \sum_{\lambda \in I} P_\lambda(z) f^{j_0} (f')^{j_1} \cdots (f^{(n)})^{j_n} = 0, \tag{2.17}$$

where I is a finite set of multi-indices $(j_0, \ldots, j_n) = \lambda$ and P_λ is polynomial for all $\lambda \in I$.

Take the terms of (2.17) with maximal degree, and among them choose terms with maximal weight. Suppose that K is the set of all these indices. Now assume that $\sum_{k \in K} P_k(z) \not\equiv 0$, and that f is an entire solution of (2.17). Then we know by [12], that

$$\rho(f) \leq \max(2d, d + 1),$$

where $d = \max_{\lambda \in I} \deg(P_\lambda)$. An upper bound for the order of growth of a meromorphic solution of an algebraic differential equation has been obtained in [6]. As mentioned earlier, we consider the case when (2.17) is homogeneous. In this special case we obtain an upper and lower bound for the order of growth of entire solutions of (2.17), by using a new method. To this end, recall the notations for the degree and the weight of an algebraic differential equation, and add the following

Definition 2.4.1. Let $P_\lambda(z)$ be a polynomial associated to each $\lambda \in I$, where I is a finite set of multi-indices, and let f be a meromorphic function of finite order $\rho(f)$. Then denote,

$$\alpha_f(\lambda) := \deg(P_\lambda) + w(\lambda)(\rho(f) - 1).$$

Now we may clearly choose $\lambda_1, \lambda_2, \ldots, \lambda_{|I|} \in I$, such that

$$\alpha_f(\lambda_1) \leq \alpha_f(\lambda_2) \leq \cdots \leq \alpha_f(\lambda_{|I|}).$$

Note that we use the index j instead of λ_j in subscripts, for convenience of notation. The following four theorems investigate the growth of transcendental entire solutions of homogeneous algebraic differential equations. We start with an upper bound estimate for the order of growth.

Theorem 2.4.2. *Let f be a transcendental entire solution of finite order of a homogeneous algebraic differential equation*

$$P(z, f) := \sum_{\lambda \in I} P_\lambda(z) f^{j_0} (f')^{j_1} \cdots (f^{(n)})^{j_n} = 0, \tag{2.18}$$

with polynomial coefficients, where $\lambda = (j_0, \ldots, j_n)$. Take $\lambda_q \in I$, and choose $\lambda_M \in I \setminus \{\lambda_q\}$ such that

$$\alpha_f(\lambda_M) = \max_{\lambda \in I \setminus \{\lambda_q\}} \alpha_f(\lambda).$$

If $w(\lambda_q) > w(\lambda_M)$, then

$$\rho(f) \leq 1 + \frac{\deg(P_M) - \deg(P_q)}{w(\lambda_q) - w(\lambda_M)}.$$

Proof. Denote $\lambda_q = (k_0, \ldots, k_n)$. Since the equation (2.18) is homogeneous, say of degree d, we may write it in the form

$$P_q(z) \frac{f^{k_0} (f')^{k_1} \cdots (f^{(n)})^{k_n}}{f^d} = - \sum_{\lambda \in I \setminus \{\lambda_q\}} P_\lambda(z) \frac{f^{j_0} (f')^{j_1} \cdots (f^{(n)})^{j_n}}{f^d}. \tag{2.19}$$

Hence by applying Theorem 1.1.2 for the left hand side of (2.19), and Theorem 1.1.3 for the right hand side, we have

$$\begin{aligned} r^{\deg(P_q)}\left(\frac{\nu(r,f)}{r}\right)^{w(\lambda_q)}|1+o(1)| &\leq \sum_{\lambda\in I\setminus\{\lambda_q\}} r^{\deg(P_\lambda)+w(\lambda)(\rho(f)-1)+\varepsilon''} \\ &\leq r^{\deg(P_M)+w(\lambda_M)(\rho(f)-1)+\varepsilon'}, \end{aligned}$$

which holds for all r large enough such that $r=|z|\notin F$, where $F\subset\mathbb{R}_+$ is a set of finite logarithmic measure, and where z is chosen so that $|f(z)|=M(r,f)$. Moreover, $\varepsilon'',\varepsilon'>0$ are arbitrarily small. Thus by Lemma 1.2.2,

$$\nu(r,f)^{w(\lambda_q)}|1+o(1)| \leq r^{\deg(P_M)-\deg(P_q)+w(\lambda_M)(\rho(f)-1)+w(\lambda_q)+\varepsilon},$$

where $\varepsilon>0$, holds for all r sufficiently large. Hence,

$$\begin{aligned} w(\lambda_q)\,\rho(f) &= \limsup_{r\to\infty}\frac{w(\lambda_q)\log^+\nu(r,f)}{\log r} \\ &\leq \deg(P_M)-\deg(P_q)+w(\lambda_M)(\rho(f)-1)+w(\lambda_q), \end{aligned}$$

and so

$$\rho(f)\leq 1+\frac{\deg(P_M)-\deg(P_q)}{w(\lambda_q)-w(\lambda_M)},$$

since $w(\lambda_q)>w(\lambda_M)$. □

Similarly, the following theorem gives us a lower bound for the order of growth.

Theorem 2.4.3. *Let f be a transcendental entire solution of finite order of a homogeneous algebraic differential equation* (2.18). *Take $\lambda_q\in I$, and choose $\lambda_M\in I\setminus\{\lambda_q\}$ such that*

$$\alpha_f(\lambda_M)=\max_{\lambda\in I\setminus\{\lambda_q\}}\alpha_f(\lambda).$$

If $w(\lambda_M)>w(\lambda_q)$, then

$$\rho(f)\geq 1+\frac{\deg(P_M)-\deg(P_q)}{w(\lambda_q)-w(\lambda_M)}.$$

Proof. By completely similar reasoning as in the proof of Theorem 2.4.2, we obtain

$$w(\lambda_q)\,\rho(f)\leq \deg(P_M)-\deg(P_q)+w(\lambda_M)(\rho(f)-1)+w(\lambda_q),$$

and so

$$\rho(f)\geq 1+\frac{\deg(P_M)-\deg(P_q)}{w(\lambda_q)-w(\lambda_M)},$$

since $w(\lambda_q) < w(\lambda_M)$. □

The following theorem gives us a connection between the coefficients and the weights of the terms of the equation (2.18).

Theorem 2.4.4. *Let f be a transcendental entire solution of a homogeneous algebraic differential equation* (2.18). *If $w(\lambda_j) = w(\lambda_k)$ for all $j, k = 1, \ldots, |I|$, then*

$$\sum_{\lambda \in I} P_\lambda(z) = 0.$$

Proof. Denote by w the weight of any term of (2.18). Then by applying Theorem 1.1.2 for all terms of (2.17), we have

$$\left(\sum_{\lambda \in I} P_\lambda(z) \right) \left(\frac{\nu(r, f)}{z} \right)^w (1 + o(1)) = 0.$$

Now suppose that $\sum_{\lambda \in I} P_\lambda(z) \not\equiv 0$. Then it follows that $\nu(r, f) \equiv 0$, which contradicts the transcendency of f. The assertion follows. □

In some cases, as the following theorem demonstrates, we are even able to determine the order of growth accurately.

Theorem 2.4.5. *Let f be a transcendental entire finite order solution of* (2.18). *If $\alpha_f(\lambda_1) = \alpha_f(\lambda_{|I|})$, and there exist $j, k = 1, \ldots, |I|$ such that $w(\lambda_j) \neq w(\lambda_k)$, then*

$$\rho(f) = 1 + \frac{\deg(P_k) - \deg(P_j)}{w(\lambda_j) - w(\lambda_k)}.$$

Proof. First of all, clearly $\alpha_f(\lambda_j) = \alpha_f(\lambda_k)$. Hence by similar reasoning as in the proof of Theorem 2.4.2, we obtain

$$w(\lambda_j)\, \rho(f) \leq \deg(P_k) - \deg(P_j) + w(\lambda_k)\, (\rho(f) - 1) + w(\lambda_j), \tag{2.20}$$

and by leaving only the term corresponding to index λ_k to the left hand side in the equation (2.19), and again, by applying Theorem 1.1.2 for the left hand side, and Theorem 1.1.3 for the right hand side of the same equation, we have

$$w(\lambda_k)\, \rho(f) \leq \deg(P_j) - \deg(P_k) + w(\lambda_j)\, (\rho(f) - 1) + w(\lambda_k). \tag{2.21}$$

By combining inequalities (2.20) and (2.21), we have our assertion. □

In the following example, we calculate the order of growth of a entire transcendental solution of a differential equation.

Example 2.4.6. Return back to the differential equation

$$ff'' - f^2 - 2f'^2 = 0, \tag{2.22}$$

which was introduced in Example 2.1.2, and assume that it has a transcendental entire solution ϕ. Then, by [12], $\rho(\phi) \leq 1$. Choose $-f^2$ to correspond to the index λ_q in Theorem 2.4.3. Then ff'' corresponds to λ_M, and so

$$\rho(\phi) \geq 1 + \frac{\deg(1) - \deg(-1)}{w(f^2) - w(ff'')} = 1.$$

Hence we conclude that $\rho(\phi) = 1$ for an arbitrary entire solution of (2.22).

Note that, in addition to a meromorphic solution $\psi(z) := (\sin z)^{-1}$, equation (2.22) has a family of entire solutions $f_C(z) = Ce^{iz}$ and $1/f_C(z)$, $C \in \mathbb{C}\setminus\{0\}$. It is immediately seen that $\rho(f_C) = \rho(1/f_C) = 1$ for all $C \in \mathbb{C}\setminus\{0\}$.

2.5. Hypertranscendental functions

Bank and Kaufman proved in [2] that the Gamma function Γ can not satisfy any algebraic differential equation over $\mathcal{L}_0$. An interesting question, which still remains open in general, is whether the zeros of Γ alone define its behaviour with respect to algebraic differential equations. In other words, does $\Gamma \notin \mathcal{A}(\mathcal{L})$ imply $\Gamma e^h \notin \mathcal{A}(\mathcal{L})$ for an entire function h? In this section, we consider the following, a more general

Question 2.5.1. Does $F \notin \mathcal{A}(\mathcal{L})$ imply $F \cdot (e^h + a) \notin \mathcal{A}(\mathcal{L})$ for an entire function h, where F is an arbitrary hypertranscendental function, and a is a small function with respect to F?

We need the following lemma in order to answer this question.

Lemma 2.5.2. *Let h be an entire function, and let $\mathcal{L}$ be a differential field. Then $e^h \in \mathcal{A}(\mathcal{L})$ if and only if $h \in \mathcal{A}(\mathcal{L})$.*

Proof. Assume first that $h \in \mathcal{A}(\mathcal{L})$. Then, since e^h satisfies

$$y' - h'y = 0,$$

$e^h \in \mathcal{A}(\mathcal{A}(\mathcal{L}))$. Hence by Theorem 1.3.1, $e^h \in \mathcal{A}(\mathcal{L})$.

Suppose now that $e^h \in \mathcal{A}(\mathcal{L})$. Then also

$$h' = \frac{(e^h)'}{e^h} \in \mathcal{A}(\mathcal{L}),$$

and so $h \in \mathcal{A}(\mathcal{L})$. □

The following example demonstrates that the answer to Question 2.5.1, in general, is negative.

Example 2.5.3. Take $h = -1/\Gamma$. Then clearly $e^{1/\Gamma}e^h \in \mathcal{A}(\mathcal{L})$ for all differential fields $\mathcal{L}$, but $e^{1/\Gamma}$ is hypertranscendental over $\mathcal{L}_0$ by Lemma 2.5.2.

However, we are able to give a positive answer to Question 2.5.1 by giving certain restrictions to a, h and F. For this we need the following corollary of Theorem 2.1.1.

Corollary 2.5.4. *Let f be a meromorphic transcendental function, let $\mathcal{L} \subset \mathcal{L}_f$ be a differential field, and let g be a meromorphic function hypertranscendental over $\mathcal{L}$. If g is a deficient function with respect to f, then f is hypertranscendental over $\mathcal{L}$ as well.*

Proof. Assume conversely that f is differentially algebraic over $\mathcal{L}$ with the minimal polynomial $M(z, f)$. Then since $\delta(g, f) > 0$, we have by Theorem 2.1.1 that $M(z, g) \equiv 0$. This contradicts the fact that g is hypertranscental over $\mathcal{L}$. Hence also f is hypertranscental over $\mathcal{L}$. □

The following lemma helps us to deal with defect relations.

Lemma 2.5.5. *Let f be a transcendental meromorphic function, and let $a, g \in \mathcal{L}_{f,0}$. Then*

1. $\delta(g + a, f) = \delta(a, f - g)$,
2. $\delta(0, gf) = \delta(0, f)$, *when* $g \not\equiv 0$.

Proof. 1. By the definition of δ,

$$\delta(g+a, f) = \liminf_{r\to\infty} \frac{m\left(r, \frac{1}{f-g-a}\right)}{T(r, f)} = \liminf_{r\to\infty} \frac{m\left(r, \frac{1}{f-g-a}\right)}{T(r, f-g) + o(T(r, f))} = \delta(a, f-g).$$

2. To prove the second equality, we first observe that

$$\delta(0, gf) \leq \liminf_{r\to\infty} \left(\frac{m\left(r, \frac{1}{f}\right)}{T(r, f)} + \frac{m\left(r, \frac{1}{g}\right)}{T(r, f)} \right) = \liminf_{r\to\infty} \frac{m\left(r, \frac{1}{f}\right)}{T(r, f)} = \delta(0, f).$$

On the other hand,

$$\begin{aligned} \delta(0, f) &= \liminf_{r\to\infty} \frac{m\left(r, \frac{1}{fgg^{-1}}\right)}{T(r, fgg^{-1})} \\ &\leq \liminf_{r\to\infty} \left(\frac{m\left(r, \frac{1}{gf}\right)}{T(r, gf) + o(T(r, f))} + \frac{m(r, g)}{T(r, gf) + o(T(r, f))} \right) \\ &= \delta(0, gf). \end{aligned}$$

By combining the two above inequalities, we have the assertion. □

Remark. The assertion of Lemma 2.5.5 is also true, if we replace δ by ϑ or Θ. This can be easily proven by using the above method.

We are now ready to prove a useful theorem, which helps us with the construction of some hypertranscendental functions.

Theorem 2.5.6. *Let G be a meromorphic function, and let F be a hypertranscendental meromorphic function over $\mathcal{L} \subset \mathcal{L}_G$ such that $T(r,F) = o(T(r,G))$. If there exists $a \in \mathcal{L}$ such that $T(r,a) = o(T(r,G))$, and*

1. *$\delta(a, G+F) > 0$, then G is hypertranscendental over $\mathcal{L}$,*
2. *$\delta(a, G) > 0$, where $a \not\equiv 0$, then FG is hypertranscendental over $\mathcal{L}$.*

Proof. 1. Since F is hypertranscendental over $\mathcal{L}$, clearly also $-F+a$ is hypertranscendental over $\mathcal{L}$. Since by Lemma 2.5.5, $\delta(-F+a, G) = \delta(a, G+F) > 0$, it follows by Corollary 2.5.4 that G is hypertranscendental over $\mathcal{L}$.

2. We have by Lemma 2.5.5,

$$0 < \delta(a, G) = \delta(0, G-a) = \delta(0, FG - aF) = \delta(aF, FG).$$

Since F is hypertranscendental over $\mathcal{L}$, also aF is hypertranscendental over $\mathcal{L}$. Hence, also FG is hypertranscendental over $\mathcal{L}$ by Corollary 2.5.4. □

In order to state the main result of this section, we need some additional notation. First, if $\mathcal{L}_1$ and $\mathcal{L}_2$ are differential fields, then

$$e^{\mathcal{L}_1} := \{e^f : f \in \mathcal{L}_1\}$$

and

$$\mathcal{L}_1\mathcal{L}_2 := \{fg : f \in \mathcal{L}_1, \quad g \in \mathcal{L}_2\}.$$

Now we are ready for the following

Definition 2.5.7. Let $\mathcal{L}$ be a differential field, let $\mathcal{E}$ denote the ring of entire functions, and let $\mathcal{L}(\mathcal{E}) := \mathcal{L} \cap \mathcal{E}$. Then

$$\mathrm{E}_n^{\mathcal{L}(\mathcal{E})} := \mathcal{L}(\mathcal{E})e^{\mathcal{L}(\mathcal{E})e^{\cdot^{\cdot^{\cdot \mathcal{L}(\mathcal{E})e^{\mathcal{L}(\mathcal{E})}}}}},$$

where the total number of $\mathcal{L}(\mathcal{E})$'s on the right hand side equals to n. In other words, $\mathrm{E}_1^{\mathcal{L}(\mathcal{E})} = \mathcal{L}(\mathcal{E})$, $\mathrm{E}_2^{\mathcal{L}(\mathcal{E})} = \mathcal{L}(\mathcal{E})e^{\mathcal{L}(\mathcal{E})}$, etc.

Theorem 2.5.8. *Let F be a hypertranscendental meromorphic function over $\mathcal{L}$, let $a \in \mathcal{L} \subset \mathcal{L}_F$, and let h be an entire function. If $h \in E_n^{\mathcal{L}(\mathcal{E})}$ for some $n \in \mathbb{N}$, then Fe^h is hypertranscendental over $\mathcal{L}$. On the other hand, if h is any entire function, and if either*

(i) F is of finite order, or

(ii) F is of infinite order, and $\max(T(r,F),T(r,a)) = o(T(r,Fe^h))$,

then the following functions are hypertranscendental over $\mathcal{L}$:

1) $F \cdot (e^h + a)$, *whenever* $a \not\equiv 0$,

2) $F + ae^h$.

Proof. We start with the hypertranscendency of Fe^h. We first prove that e^h is differentially algebraic over $\mathcal{L}$, whenever $h \in \mathrm{E}_n^{\mathcal{L}(\mathcal{E})}$.

Assume first that $n = 1$. Then e^h is differentially algebraic over $\mathcal{L}$ by Lemma 2.5.2.

We use induction to prove the remaining cases. Suppose that e^h is differentially algebraic over $\mathcal{L}$, whenever $h \in \mathrm{E}_k^{\mathcal{L}(\mathcal{E})}$ for any $k = 1, \ldots, n-1$. Then, if $h \in \mathrm{E}_n^{\mathcal{L}(\mathcal{E})}$,

$$e^h = e^{h_1 e^{h_2}}$$

for some entire functions $h_1, h_2 \in \mathcal{A}(\mathcal{L})$. Then by Lemma 2.5.2, $e^{h_2} \in \mathcal{A}(\mathcal{L})$. Hence also $h_1 e^{h_2} \in \mathcal{A}(\mathcal{L})$, and so, by Lemma 2.5.2 again, $e^h \in \mathcal{A}(\mathcal{L})$. Thus by the induction principle, $e^h \in \mathcal{A}(\mathcal{L})$ whenever $h \in \mathrm{E}_n^{\mathcal{L}(\mathcal{E})}$.

Suppose now that Fe^h is differentially algebraic over $\mathcal{L}$. Since e^h is differentially algebraic over $\mathcal{L}$, also e^{-h} differentially algebraic over the same field. Hence,

$$F = (Fe^h)e^{-h}$$

is also differentially algebraic over $\mathcal{L}$. This contradicts our assumption, and so Fe^h must be hypertranscendental over $\mathcal{L}$.

1. We now prove the hypertranscendency of $F \cdot (e^h + a)$.

(i) We start with the case $\rho(F) < \infty$. Suppose first that $\rho(e^h) < \infty$. Then clearly h is a polynomial, say of order q. Assume that $F \cdot (e^h + a)$ is differentially algebraic over $\mathcal{L}$. Now, e^h satisfies

$$\left(\frac{y'}{y}\right)^{(q)} = 0,$$

which is equivalent to an algebraic differential equation for y with integer coefficients. Since integers are contained in $\mathcal{L}$, e^h is differentially algebraic over $\mathcal{L}$. Thus also $1/(e^h + a)$ is differentially algebraic over $\mathcal{L}$, and so

$$F = F \cdot (e^h + a) \cdot \frac{1}{e^h + a}$$

is differentially algebraic over $\mathcal{L}$, which is a contradiction. Hence $F \cdot (e^h + a)$ must be hypertranscendental over $\mathcal{L}$.

Assume now that e^h is of infinite order. Since zero and infinity are Borel exceptional values for e^h, it is of regular growth. Hence, $\mu(e^h) = \infty$, and so $T(r, F) = o(T(r, e^h))$. Suppose now that $\alpha > 1$ and $\varepsilon > 0$ are real numbers. Then by Lemma 1.2.1, $T(r, a) = o(T(\alpha r, F)) = o(T(r, e^h))$, since F is of finite order. This implies that

$$\begin{aligned}\mu(e^h + a) &= \liminf_{r\to\infty} \frac{\log T(r, e^h + a)}{\log r} \\ &\geq \liminf_{r\to\infty} \frac{\log \left(T(r, e^h) - T(r, a)\right)}{\log r} \\ &\geq \liminf_{r\to\infty} \frac{\log \left((1-\varepsilon) T(r, e^h)\right)}{\log r} \\ &= \liminf_{r\to\infty} \left(\frac{\log(1-\varepsilon)}{\log r} + \frac{\log T(r, e^h)}{\log r} \right) = \infty.\end{aligned}$$

Hence also $\rho(e^h + a) = \infty$. Now fix β such that $\rho(F) < \beta < \rho\left(e^h + a\right)$. Since $e^h + a$ is of regular growth, there exists $r_0 > 0$ such that $T(r, e^h + a) > r^\beta$ for all $r > r_0$. Thus

$$T(r, F) = o\left(T(r, e^h + a)\right). \tag{2.23}$$

Since by Lemma 2.5.5,

$$\delta(a, e^h + a) = \delta(0, e^h) = 1 \tag{2.24}$$

we have by Theorem 2.5.6, that $F \cdot (e^h + a)$ is hypertranscendental over $\mathcal{L}$.

(ii) Assume now that $\rho(F) = \infty$ and that $\max(T(r, F), T(r, a)) = o(T(r, Fe^h))$. Then by Lemma 2.5.5, the equation (2.24) holds. Hence the assertion follows from Theorem 2.5.6.

2. Consider now the function $F + ae^h$. Since the case $a \equiv 0$ is trivial, we may assume that $a \not\equiv 0$.

(i) Consider first the case $\rho(F) < \infty$. Suppose that $\rho(e^h) < \infty$. Assume conversely that $F + ae^h$ is differentially algebraic over $\mathcal{L}$. Similarly as in the early part of this proof, we can see that e^h is differentially algebraic over $\mathcal{L}$. Thus also

$$F = \left(F + ae^h\right) - ae^h$$

is differentially algebraic over $\mathcal{L}$, which is a contradiction. Hence $F + ae^h$ must be hypertranscendental over $\mathcal{L}$.

Suppose now that $\rho(e^h) = \infty$, and let $\varepsilon > 0$. Since $T(r, F) = o(T(r, e^h))$ and $T(r, a) = o(T(r, e^h))$, we have

$$\begin{aligned}\mu(F + ae^h) &= \liminf_{r\to\infty} \frac{\log T(r, F + ae^h)}{\log r} \\ &\geq \liminf_{r\to\infty} \frac{\log \big(T(r, e^h) - T(r, F) - T(r, a)\big)}{\log r} \\ &\geq \liminf_{r\to\infty} \frac{\log \big((1-\varepsilon)T(r, e^h)\big)}{\log r} \\ &= \liminf_{r\to\infty} \left(\frac{\log(1-\varepsilon)}{\log r} + \frac{\log T(r, e^h)}{\log r}\right) = \infty.\end{aligned}$$

Hence also $F + ae^h$ is of regular growth, and so $T(r, F) = o(T(r, F + ae^h))$. Now since,

$$N\left(r, \frac{1}{F + ae^h - F}\right) = N\left(r, \frac{1}{ae^h}\right) = o(T(r, ae^h)), \tag{2.25}$$

we have

$$1 = \delta(0, F + ae^h - F) = \delta(F, F + ae^h), \tag{2.26}$$

by Lemma 2.5.5. Hence $F + ae^h$ is hypertranscendental with respect to $\mathcal{L}$ by Corollary 2.5.4.

(ii) To prove the infinite order case, we note that since we can prove that $T(r, F) = o(T(r, F + ae^h))$ holds similarly as in the above proof, (2.26) holds in this case. Since the equation (2.25) also holds, F is a deficient function with respect to $F + ae^h$. Hence we have our assertion by Corollary 2.5.4. □

The following example illustrates the use of Theorem 2.5.8.

Example 2.5.9. We know by [2] that Γ is hypertranscendental over $\mathcal{L}_0$. Define inductively, $\exp_1(z) := e^z$ and $\exp_{n+1}(z) := \exp(\exp_n(z))$, $n \in \mathbb{N}$. We know by Lemma 2.5.2 that $\exp_n(z)$ is differentially algebraic over $\mathcal{L}_0$ for all $n \in \mathbb{N}$. On the other hand, $\Gamma \exp_n(z)$ is hypertranscendental over $\mathcal{L}_0$ by Theorem 2.5.8. Note that by multiplying the function $\exp_n(z)$ with the Gamma function we do not affect its growth significantly. However, the behavior of $\exp_n(z)$ with respect to algebraic differential equations changes radically in this process from differentially algebraic to hypertranscendental.

The following corollary follows easily from Theorem 2.5.8.

Corollary 2.5.10. *Let F be a hypertranscendental meromorphic function over $\mathcal{L}$, such that $\rho(F) < \infty$. Let*

$$G(z) := \sum_{n=1}^{k} g_n(z),$$

where $k \in \mathbb{N}$ and $g_n(z)$ are entire functions of finite order with no zeros. Then $F+G$ is hypertranscendental over $\mathcal{L}$.

Proof. Assertion follows easily by inductive use of Theorem 2.5.8 in the finite order case. □

Example 2.5.11. Since Γ is hypertranscendental over $\mathcal{L}_0$ and

$$\sin z = \frac{1}{2i}e^{iz} - \frac{1}{2i}e^{-iz},$$

Corollary 2.5.10 implies that also $\Gamma(z) + \sin(z)$ is hypertranscendental over $\mathcal{L}_0$.

3. Algebraic differential equations of the first order

In this section, we restrict our considerations to the solutions of first order differential equations. We mainly concern ourselves with the equations of the generalized Malmquist type, and the equations connected with them, like the Riccati differential equation.

3.1. Linear homogeneous case

As an introduction for the first order differential equations, we consider briefly the trivial case of linear homogeneous first order differential equations, for reasons that will become clear later on in this section. So, let now f be a meromorphic function such that

$$\Phi(f) := \frac{a(z)f + b(z)}{c(z)f + d(z)}, \tag{3.1}$$

is a solution of the differential equation

$$u' + A(z)u = 0 \tag{3.2}$$

for a suitable meromorphic coefficient A, such that $e^{-\int A(z)dz}$ exists and is meromorphic. We assume further that a, b, c, d are small functions with respect to f, such that $a(z)d(z) \not\equiv b(z)c(z)$. By (3.1) and (3.2),

$$\Phi(f) = Ce^{-\int A(z)dz},$$

where $C \in \mathbb{C}$, and so

$$f = \frac{b(z) - d(z)Ce^{-\int A(z)dz}}{c(z)Ce^{-\int A(z)dz} - a(z)}.$$

Hence f is explicitly solvable, and this case is completely known.

3.2. Equation of generalized Malmquist type

Consider first the differential equation

$$u' = R(z, u), \tag{3.3}$$

where R is rational in both arguments. The celebrated Malmquist theorem states that all differential equations of the form (3.3) admitting transcendental meromorphic solutions, must reduce into a Riccati differential equation

$$u' = a_0(z) + a_1(z)u + a_2(z)u^2 \tag{3.4}$$

with rational coefficients. Similarly, N. Steinmetz characterized all cases where transcendental meromorphic solutions of the equation

$$(u')^n = R(z, u) \tag{3.5}$$

may appear [33]. His result is stated as follows:

Theorem 3.2.1 (Steinmetz). *Let $R(z, u)$ be rational in both of its arguments. If (3.5) admits a transcendental meromorphic solution, then after a suitable Möbius transformation*

$$u = \frac{\alpha\nu + \beta}{\gamma\nu + \delta},$$

where $\alpha\delta - \beta\gamma \neq 0$, (3.5) reduces into one of the following types:

$$\begin{aligned}
\nu' &= a(z) + b(z)\nu + c(z)\nu^2, && (3.6)\\
(\nu')^2 &= a(z)(\nu - b(z))^2(\nu - \tau_1)(\nu - \tau_2), && (3.7)\\
(\nu')^2 &= a(z)(\nu - \tau_1)(\nu - \tau_2)(\nu - \tau_3)(\nu - \tau_4), && (3.8)\\
(\nu')^3 &= a(z)(\nu - \tau_1)^2(\nu - \tau_2)^2(\nu - \tau_3)^2, && (3.9)\\
(\nu')^4 &= a(z)(\nu - \tau_1)^2(\nu - \tau_2)^3(\nu - \tau_3)^3, && (3.10)\\
(\nu')^6 &= a(z)(\nu - \tau_1)^3(\nu - \tau_2)^4(\nu - \tau_3)^5, && (3.11)
\end{aligned}$$

where $\tau_1, \tau_2, \tau_3, \tau_4$ are complex constants, and the coefficients $a(z), b(z), c(z)$ are rational functions. Moreover, $a(z) \not\equiv 0$ in (3.7)–(3.11).

The next theorem shows that if a solution of (3.5) has, in a sense, two deficient slowly moving targets, then it satisfies a linear homogeneous first order differential equation.

Theorem 3.2.2. *Let u be a solution of (3.5), and let $g_1, g_2 \in \mathcal{L}_u$ such that*

$$\overline{N}\left(r, \frac{1}{u - g_j}\right) = S(r, u) \tag{3.12}$$

for $j = 1, 2$, and such that at least one g_j is non-constant. Then:

a) If $g_2 \neq \frac{\alpha}{\gamma}$, after a suitable Möbius transformation

$$u = \frac{(\gamma g_1 g_2 - \alpha g_2)f - (\gamma g_1 g_2 - \alpha g_1)}{(\gamma g_1 - \alpha)f - (\gamma g_2 - \alpha)}, \tag{3.13}$$

where $\alpha\delta - \beta\gamma \neq 0$, (3.5) reduces into, separately in each of the six cases (3.6)–(3.11),

$$B_k(z)f' = A_k(z)f, \tag{3.14}$$

where A_k and B_k, $k = 1, \ldots, 6$, are differential polynomials in g_1,g_2 with rational coefficients.

b) If $g_2 = \frac{\alpha}{\gamma}$, after a suitable Möbius transformation

$$u = \frac{\alpha(\gamma g_1 - \alpha)f + (\beta\gamma - \alpha\delta)g_1}{\gamma(\gamma g_1 - \alpha)f + \beta\gamma - \alpha\delta} \tag{3.15}$$

where $\alpha\delta - \beta\gamma \neq 0$, (3.5) reduces into

$$f' = b(z)f,$$

where $b(z)$ is a rational function.

Remark. Since most of the expressions A_k and B_k are very complicated, their explicit forms are given in Appendix A, see (A.1)–(A.6) and (A.7)–(A.12).

Proof. We know by Theorem 3.2.1 that

$$\nu := \frac{\beta - \delta u}{\gamma u - \alpha}, \tag{3.16}$$

where $\alpha\delta - \beta\gamma \neq 0$, satisfies one of the equations (3.6)–(3.11). We may assume that g_1 is non-constant, since otherwise we can interchange g_1 and g_2. Assume first that $g_2 \neq \alpha/\gamma$. Then, since (3.12) holds, we have by (3.16)

$$\overline{N}\left(r, \frac{1}{\nu - h_j}\right) = S(r, u) = S(r, \nu), \tag{3.17}$$

where

$$h_j := \frac{\beta - \delta g_j}{\gamma g_j - \alpha} \tag{3.18}$$

for $j = 1, 2$. Now, by defining

$$f := \frac{\nu - h_1}{\nu - h_2}, \tag{3.19}$$

we obtain the Möbius transformation (3.13) by solving u from the equation (3.19), using (3.16).

We next demonstrate that f satisfies (3.14) with a pair of coefficients A_k, B_k, $k = 1, \dots, 6$. Suppose first that ν satisfies (3.6). Substitute

$$\nu = \frac{h_2 f - h_1}{f - 1} \tag{3.20}$$

into (3.6) to obtain

$$\begin{aligned}(h_2' - a - bh_2 - ch_2^2)f^2 &- (2a + h_1' + h_2' + bh_1 + bh_2 + 2ch_1h_2)f \\ &+ (h_1 - h_2)f' + h_1' - a - bh_1 - ch_1^2 = 0.\end{aligned} \tag{3.21}$$

Now, by (3.17) and (3.19), $\max(\overline{N}(r, f), \overline{N}(r, 1/f)) = S(r, f)$, and so by Lemma 2.3.2, each homogeneous component of (3.21) must vanish identically. Hence,

$$(h_1 - h_2)f' + (2a + h_1' + h_2' + bh_1 + bh_2 + 2ch_1h_2)f = 0,$$

and h_1, h_2 must both satisfy equation (3.6). Thus we have the case A_1, B_1.

Assume now that ν satisfies (3.7). Similarly by substituting (3.20) into (3.7), we obtain

$$\begin{aligned}((h_2')^2 - a(h_2 - b)^2(h_2 - c_1)(h_2 - c_2))f^4 &+ \Omega(z, f, f') \\ + \Big(- 4ah_2h_1^3 &+ 2h_2'h_1' - 2ah_1^3b + 2ah_1bc_1c_2 \\ + ab^2c_1h_1 &- 2ab^2c_1c_2 + ab^2h_1c_2 + 2ah_1^4 + 2ah_2bc_1c_2 + ab^2h_2c_2 \\ + ab^2c_1h_2 &- 2ah_2h_1c_1c_2 - 4ah_2bh_1c_2 - 4ah_2bc_1h_1 - 2ab^2h_2h_1 \\ + 3ah_2h_1^2c_1 &+ 6ah_2bh_1^2 + 3ah_2h_1^2c_2 - ah_1^3c_2 - ah_1^3c_1\Big) f \\ + h_1'(h_1 - h_2)f' &+ (h_1')^2 - a(h_1 - b)^2(h_1 - c_1)(h_1 - c_2) = 0,\end{aligned} \tag{3.22}$$

where $\Omega(z, f, f')$ is a differential polynomial over $\mathcal{L}_f$ having no terms of degree zero, one or four. Again, by (3.17), and by Lemma 2.3.2, each homogeneous component of (3.22) must vanish identically. Hence h_1, h_2 must satisfy (3.7), and

$$B_2(z)f' = A_2(z)f.$$

Note that since g_1 was assumed to be non-constant, so is h_1, and so $B_2 \not\equiv 0$, which implies that also $A_2 \not\equiv 0$.

By repeating the above process for each equation (3.8)–(3.11), we obtain

$$B_k(z)f' = A_k(z)f, \tag{3.23}$$

where $k = 3, \dots, 6$. Also both h_1 and h_2 must satisfy one of the equations (3.8)–(3.11). We omit the details, since the idea is exactly the same as above.

The calculations in the cases $k = 2, \ldots, 6$ were done by Maple©, because of the complicated nature of the corresponding expressions.

Assume finally that $g_2 = \alpha/\gamma$. Since then $h_2 = \infty$, and $\overline{N}(r, \nu) = S(r, \nu)$, we put $f = \nu - h_1$, and we obtain the Möbius transformation (3.15) by solving u from this equation. By substituting $\nu = f + h_1$ into one of the equations (3.7)–(3.11), we obtain an equation of the form

$$af^m = \Omega_m(z, f', f),$$

where $\Omega(z, f, f')$ is a differential polynomial over $\mathcal{L}_f$ such that $d(\Omega_m) < m$. Moreover, $\max(\overline{N}(r, f), \overline{N}(r, 1/f)) = S(r, f)$, and so $a \equiv 0$ by Lemma 2.3.2, which contradicts the assertion of Theorem 3.2.1. Thus ν can not be a solution of any of the equations (3.7)–(3.11). Therefore, assume that ν satisfies (3.6), and substitute $\nu = f + h_1$ into (3.6) to obtain:

$$-c(z)f^2 + (-2c(z)h_1 - b(z))f + f' + h_1' - a(z) - c(z)h_1^2 - b(z)h_1 = 0.$$

Now, Lemma 2.3.2 implies that $c(z) \equiv 0$, and so

$$f' = b(z)f.$$

□

Remarks. 1. If the assumptions of Theorem 3.2.2 all hold, the solution of (3.5) is completely known, see Section 3.1.

2. In fact, K. Ishizaki has done the above reduction for Riccati differential equation in the case $\delta(g_j, f) = 1$, $j = 1, 2$, see [18]. This is a slightly stronger assumption than what we had in Theorem 3.2.2.

3. The above proof shows that if one of the equations (3.6)–(3.11) has a solution f such that

$$\overline{N}\left(r, \frac{1}{f - g_j}\right) = S(r, f)$$

for $g_j \in \mathcal{L}_f$, $j = 1, 2$, then g_1 and g_2 must also satisfy the same equation.

The following theorem considers the case of two constant deficient functions.

Theorem 3.2.3. *Let u be a solution of* (3.5), *and let $g_1, g_2 \in \mathbb{C}$ such that*

$$\overline{N}\left(r, \frac{1}{u - g_j}\right) = S(r, u) \tag{3.24}$$

for $j = 1, 2$. Then after a suitable Möbius transformation (3.13) *or* (3.15), (3.5) *reduces into one of the following equations:*

$$
\begin{aligned}
f' &= b(z)f, && (3.25)\\
B_1(z)f' &= A_1(z)f, && (3.26)\\
D_1(z)(f')^2 &= C_1(z)f^2, && (3.27)\\
D_2(z)(f')^2 &= C_2(z)f^2, && (3.28)\\
D_3(z)(f')^3 &= C_3(z)f^3, && (3.29)\\
D_4(z)(f')^4 &= C_4(z)f^4, && (3.30)\\
D_5(z)(f')^6 &= C_5(z)f^6, && (3.31)
\end{aligned}
$$

where C_k and D_k are rational functions, and b, A_1 and B_1 are the same as in Theorem 3.2.2.

Remark. Since again most of the expressions C_k and D_k are complicated, their explicit forms are given in Appendix B, see (B.1)–(B.5) and (B.6)–(B.10).

Proof. Firstly, the case $g_2 = \alpha/\gamma$ leads, by the Möbius transformation (3.15), to the equation (3.25) by exactly the same reasoning as in the proof of Theorem 3.2.2.

Assume now that $g_2 \neq \alpha/\gamma$. We obtain the Möbius transformation (3.13) by completely the same reasoning as in the proof of Theorem 3.2.2. We also know by Theorem 3.2.1 that

$$\nu := \frac{\beta - \delta u}{\gamma u - \alpha} \tag{3.32}$$

satisfies one of the equations (3.6)–(3.11). In addition, the functions h_1, h_2 are as in (3.18).

Suppose first that ν satisfies (3.6). This leads to the case (3.26) by completely similar reasoning as in the proof of Theorem 3.2.2.

Assume now that ν satisfies (3.7). By substituting (3.20) into (3.7), we obtain

$$
\begin{aligned}
\Omega(z,f,f') + \Big(&-4ah_2h_1c_1c_2 + 6ah_2bc_1c_2 - 8ah_2bh_1c_2 - 2ah_2^2bc_1\\
&- 4ab^2h_1h_2 + 3ah_2^2h_1c_2 - ah_2^2c_1c_2 + 3ab^2c_1h_2 + 3ah_2h_1^2c_2 + 3ab^2h_2c_2\\
&+ 6ah_2^2bh_1 - 2ah_2^2bc_2 + 3ah_1^2c_1h_2 + 6ah_1^2bh_2 + 3ah_2^2h_1c_1 - 8ah_1bc_1h_2\\
&- 6ah_2^2h_1^2 - ab^2h_2^2 - 2ah_1^2bc_1 - ab^2h_1^2 - ah_1^2c_1c_2 - 2ah_1^2bc_2 + 3ab^2h_1c_2\\
&- 6ab^2c_1c_2 + 3ab^2c_1h_1 + 6ah_1bc_1c_2\Big)f^2 + (h_2-h_1)^2(f')^2 = 0,
\end{aligned} \tag{3.33}
$$

where $\Omega(z,f,f')$ is a differential polynomial over $\mathbb{C}(z)$ having no terms of degree two. Again, by (3.17), (3.19) and Lemma 2.3.2, each homogeneous component of (3.33) must vanish identically. Hence,

$$D_1(z)(f')^2 = C_1(z)f^2.$$

Moreover since clearly $D_1 \not\equiv 0$, also $C_1 \not\equiv 0$.

By repeating the above process for each equation (3.8)–(3.11), we obtain the equations (3.28)–(3.31). We omit the details, since the idea is exactly the same as above. The calculations for C_k, D_k, $k = 1, \ldots, 5$ were done by Maple©, because of the complicated nature of the corresponding expressions. □

Example 3.2.4. Suppose that f is a solution of Riccati differential equation (3.4) with constant coefficient a_2. Suppose further that f has two exceptional values $g_1, g_2 \in \mathbb{C}$ in the sense of (3.24). Then by Theorem 3.2.3,

$$h(z) := \frac{f(z) - g_1}{f(z) - g_2} \tag{3.34}$$

satisfies

$$h' = a_2(g_1 - g_2)h.$$

Hence,

$$h = c_0 e^{a_2(g_1 - g_2)z},$$

where $c_0 \in \mathbb{C}$. Therefore, by (3.34)

$$f = \frac{g_2 c_0 e^{a_2(g_1-g_2)z} - g_1}{c_0 e^{a_2(g_1-g_2)z} - 1}.$$

If we assume further that $g_1 = -g_2$, and take c_0 such that $|c_0| = 1$, then we have

$$f(z) = \frac{i g_1}{\tan(a_2 g_1 z + d_0)},$$

where d_0 is such that $\exp(2 i a_2 g_1 d_0) = c_0$.

3.3. Riccati differential equation

Since, of the equations (3.6)–(3.11), the Riccati differential equation is of particular interest due to the classical Malmquist's theorem mentioned earlier, we devote this section to study the value distribution of its meromorphic solutions more carefully.

Let R be a rational function of the form $R = P/Q$, where P and Q are polynomials with no common zeros. Then define by

$$\mathrm{di}(R) := \deg(P) - \deg(Q)$$

the *degree* of the rational function R *at infinity.*

We consider the deficiency of solutions of the Riccati differential equation

$$f' = a_0(z) + a_1(z)f + a_2(z)f^2, \quad a_0(z), a_2(z) \not\equiv 0, \tag{3.35}$$

with rational coefficients such that

$$\mathrm{di}(a_2) > \max\left(\mathrm{di}(a_0), \mathrm{di}(a_1), -2\right).$$

By multiplying out the denominators, we can write (3.35) in the form

$$b_0(z)f' = b_1(z) + b_2(z)f + b_3(z)f^2, \quad b_0(z), b_1(z), b_3(z) \not\equiv 0, \tag{3.36}$$

where b_j, $j = 0, \ldots, 3$ are polynomials such that

$$d := \deg(b_3) > \max\left(\deg(b_0) - 2, \deg(b_1), \deg(b_2)\right). \tag{3.37}$$

We know, for example by [21, p. 174], that all meromorphic solutions of the equation (3.36) are of finite order, and that all poles of the solutions, except possibly finitely many, are simple.

Lemma 3.3.1. *Let ψ be a transcendental meromorphic solution of* (3.36), *and let $n \in \mathbb{N} \cup \{0\}$. Then*

$$m(r, \psi^{(n)}) = O(\log r), \tag{3.38}$$

and

$$m\left(r, \frac{1}{\psi^{(n)} - P}\right) = O(\log r) \tag{3.39}$$

for all polynomials P.

Proof. First observe that the case $n = 0$ follows directly by Lemma 2.2.1 for (3.38). Hence, by the lemma of the logarithmic derivative,

$$m(r, \psi^{(n)}) \leq m\left(r, \frac{\psi^{(n)}}{\psi}\right) + m(r, \psi) = O(\log r).$$

Consider now (3.39) for the case $P \equiv 0$. The case $n = 0$ follows by Lemma 2.2.2, since $b_1(z) \not\equiv 0$. For the case $n = 1$ we first note that

$$m\left(r, \frac{1}{\psi'}\right) \leq m\left(r, \frac{\psi^2}{\psi'}\right) + m\left(r, \frac{1}{\psi^2}\right) = m\left(r, \frac{\psi^2}{\psi'}\right) + O(\log r). \tag{3.40}$$

By differentiating (3.36) $d = \deg(b_3)$ times, we have according to (3.37),

$$\begin{aligned}\sum_{j=0}^{d} \binom{d}{j} b_0^{(d-j)} \psi^{(j+1)} &= \sum_{j=0}^{d} \binom{d}{j} b_2^{(d-j)} \psi^{(j)} + \sum_{j=0}^{d} \binom{d}{j} b_3^{(d-j)} (\psi^2)^{(j)} \\ &= \sum_{j=1}^{d} \binom{d}{j} b_2^{(d-j)} \psi^{(j)} + c\psi^2 \\ &\quad + \sum_{j=1}^{d} \sum_{k=0}^{j} \binom{d}{j} \binom{j}{k} b_3^{(d-j)} \psi^{(j-k)} \psi^{(k)},\end{aligned}$$

where $c \in \mathbb{C}$. Thus

$$\psi^2 = Q(z, \psi),$$

where Q is a differential polynomial in ψ and its derivatives with polynomial coefficients, such that each term of Q has at least one $\psi^{(l)}$, $l \geq 1$, as a factor. Hence by (3.38), (3.40), and by the lemma of the logarithmic derivative,

$$m\left(r, \frac{1}{\psi'}\right) \leq m\left(r, \frac{\psi^2}{\psi'}\right) + O(\log r) = O(\log r).$$

Assume now that

$$m\left(r, \frac{1}{\psi^{(n)}}\right) = O(\log r) \tag{3.41}$$

has been proved for all $n = 1, \ldots, k$. By differentiating (3.36) $2k + d$ times, we have

$$(\psi^{(k)})^2 = Q(z, \psi),$$

where Q is a differential polynomial in ψ and its derivatives with polynomial coefficients, such that each term of Q has at least one $\psi^{(l)}$, $l \geq k+1$, as a factor. Hence by (3.41) ,

$$\begin{aligned} m\left(r, \frac{1}{\psi^{(k+1)}}\right) &\leq m\left(r, \frac{(\psi^{(k)})^2}{\psi^{(k+1)}}\right) + m\left(r, \frac{1}{(\psi^{(k)})^2}\right) \\ &= m\left(r, \frac{(\psi^{(k)})^2}{\psi^{(k+1)}}\right) + O(\log r) = O(\log r). \end{aligned}$$

Thus

$$m\left(r, \frac{1}{\psi^{(n)}}\right) = O(\log r) \tag{3.42}$$

for all $n \in \mathbb{N} \cup \{0\}$ by the induction principle.

Suppose now that $P \not\equiv 0$, and that $\deg(P) = p$. Then we have for all $n \in \mathbb{N} \cup \{0\}$,

$$m\left(r, \frac{1}{\psi^{(n)} - P}\right) \leq m\left(r, \frac{(\psi^{(n)} - P)^{(p+1)}}{\psi^{(n)} - P}\right) + m\left(r, \frac{1}{\psi^{(n+p+1)}}\right) = O(\log r),$$

by (3.42), and by the lemma of the logarithmic derivative. □

We can now use Lemma 3.3.1 to study the value distribution of the derivatives of the meromorphic solutions of (3.36). As the following theorems demonstrate, the distribution of values turns out to be remarkably regular.

Theorem 3.3.2. *Let ψ be a transcendental meromorphic solution of* (3.36), *and let $n \in \mathbb{N} \cup \{0\}$. Then*

$$\Delta(P, \psi^{(n)}) = \limsup_{r \to \infty} \frac{m\left(r, \frac{1}{\psi^{(n)} - P}\right)}{T(r, \psi^{(n)})} = 0$$

for all polynomials P.

Proof. The assertion follows immediately by Lemma 3.3.1. □

An immediate question is, whether or not the assertion of Theorem 3.3.2 remains valid, if we replace the polynomial P with an arbitrary rational function. The following lemma helps to find an answer to this question.

Lemma 3.3.3. *Let ψ be a transcendental meromorphic solution of* (3.36), *and let $g \in \mathbb{C}(z)$. Then $\Delta(g, \psi) = 1$ if and only if g satisfies* (3.36).

Proof. If g is not a solution of (3.36) then we have by Theorem 2.2.3 that $\Delta(g, \psi) = 0$.

Assume now that g is a solution of (3.36). Then by substituting $f = F + g$ into (3.36), we obtain an equation of the form

$$\begin{aligned} F' &= -g' + a_0 + a_1 g + a_2 g^2 + (a_1 + 2a_2 g)F + a_2 F^2 \\ &= (a_1 + 2a_2 g + a_2 F)F \end{aligned} \tag{3.43}$$

having $F = \psi - g$ as a solution. Suppose that $z_0 \in \mathbb{C}$ is such that $F(z_0) = 0$. Then the constant function zero satisfies the equation (3.43) with the initial condition $F(z_0) = 0$. Hence $F \equiv 0$ by the uniqueness theorem of differential equations. Thus $\psi \equiv g$, which is a contradiction. Hence F can not have any zeros, and so $\Delta(0, F) = \Delta(g, \psi) = 1$. □

Remark. We do not need to make the restriction (3.37) about the coefficients of (3.36) in Lemma 3.3.3, the assertion holds for all transcendental meromorphic solutions of the Riccati differential equation with rational coefficients.

The following example gives a negative answer to the question about rational functions in Theorem 3.3.2.

Example 3.3.4. Consider a differential equation

$$w' = w^2 + z^2 w + \frac{1}{4}(z^4 - 4z + 1). \tag{3.44}$$

As Ishizaki pointed out in [18], equation (3.44) has an admissible solution

$$w_0 := \frac{1}{Ce^{iz} - i} - \frac{i + z^2}{2},$$

where $C \neq 0$, and two small solutions, $P_1 := (-i - z^2)/2$ and $P_2 := (i - z^2)/2$, with respect to w_0. Now substitute $w = \alpha + u^{-1}$, $\alpha \in \mathbb{C}$, into (3.44) to obtain

$$u' = -u^2(\alpha^2 + \alpha z^2 + \frac{1}{4}(z^4 - 4z + 1)) - u(2\alpha + z^2) - 1 \tag{3.45}$$

having an admissible solution

$$f_0 := \frac{1}{w_0 - \alpha},$$

and two small solutions, $R_1 := 1/(P_1 - \alpha)$ and $R_2 := 1/(P_2 - \alpha)$, with respect to f_0. Now, by Theorem 3.3.2, $\Delta(P, f_0^{(n)}) = 0$ for all $n \in \mathbb{N} \cup \{0\}$, and for all polynomials P. Conversely, by Lemma 3.3.3, $\Delta(R_1, f_0) = \Delta(R_2, f_0) = 1$.

In order to further investigate the value distribution of ψ, we need one more definition.

Definition 3.3.5. Let f be a meromorphic function, let $a \in \mathcal{L}_f$, and let $n \in \mathbb{N} \cup \{0\}$. Then,

$$\Phi_{(n)}(f, a) := \liminf_{r\to\infty} \frac{N\left(r, \frac{1}{f^{(n)}-a}\right)}{T(r, f)}.$$

Example 3.3.6. Take $f(z) \equiv e^z$, and let $n \in \mathbb{N}$. Then $\Phi_{(n)}(f, c) = 1$ for all $c \in \mathbb{C} \setminus \{0\}$, and $\Phi_{(n)}(f, 0) = 0$.

Theorem 3.3.7. *Let ψ be a transcendental meromorphic solution of* (3.36), *and let $n \in \mathbb{N} \cup \{0\}$. Then*

$$\Phi_{(n)}(\psi, P) = n + 1$$

for all polynomials P.

Proof. First of all, note that since all poles of ψ, except at most finitely many, are simple,

$$N(r, \psi^{(n)}) = N(r, \psi) + n \cdot \overline{N}(r, \psi) = (n+1)T(r, \psi) + O(\log r).$$

Thus by Lemma 3.3.1,

$$\begin{aligned} N\left(r, \frac{1}{\psi^{(n)} - P}\right) &= T\left(r, \psi^{(n)} - P\right) - m\left(r, \frac{1}{\psi^{(n)} - P}\right) + O(1) \\ &= T\left(r, \psi^{(n)}\right) + O(\log r) \\ &= N\left(r, \psi^{(n)}\right) + O(\log r) \\ &= (n+1)T(r, \psi) + O(\log r), \end{aligned}$$

and so $\Phi_{(n)}(\psi, P) = n + 1$. □

The following theorem tells us about the growth of ψ. Its proof is included in the proof of Theorem 3.3.7.

Theorem 3.3.8. *Let ψ be a transcendental meromorphic solution of* (3.36), *and let $n \in \mathbb{N}$. Then*

$$T(r, \psi^{(n)}) = (n+1)T(r, \psi) + O(\log r).$$

Similarly to Theorem 3.3.7, we have

Theorem 3.3.9. *Let ψ be a transcendental meromorphic solution of* (3.36), *and let $n \in \mathbb{N} \cup \{0\}$. Then*

$$\Phi_e(\psi^{(n)}) = \frac{n+2}{n+1}.$$

Proof. By Lemma 3.3.1 and Theorem 3.3.8, we have

$$\Phi_e(\psi^{(n)}) = \liminf_{r\to\infty} \frac{N\left(r, \frac{1}{(\psi^{(n)})'}\right)}{T(r, \psi^{(n)})} = \liminf_{r\to\infty} \frac{(n+2)T(r, \psi) + O(\log r)}{(n+1)T(r, \psi) + O(\log r)} = \frac{n+2}{n+1}.$$

□

We also have the following version of the second main theorem for the solutions of (3.36).

Theorem 3.3.10. *Let ψ be a transcendental meromorphic solution of* (3.36), *let $P_1, \ldots, P_q$ be distinct complex polynomials, and let $n \in \mathbb{N} \cup \{0\}$. Then*

$$\sum_{j=1}^{q} m\left(r, \frac{1}{\psi^{(n)} - P_j}\right) + N_B(r, \psi^{(n)}) = (2n+2)T(r, \psi) + O(\log r).$$

Proof. Now, $N(r, \psi^{(n)}) - \overline{N}(r, \psi^{(n)}) = nN(r, \psi)$. Hence by Lemma 3.3.1, and Theorem 3.3.8

$$\begin{aligned}
\sum_{j=1}^{q} m\left(r, \frac{1}{\psi^{(n)} - P_j}\right) + N_B(r, \psi^{(n)}) &= \sum_{j=1}^{p} m\left(r, \frac{1}{\psi^{(n)} - P_j}\right) + N\left(r, \frac{1}{(\psi^{(n)})'}\right) \\
&\quad + N(r, \psi^{(n)}) - \overline{N}(r, \psi^{(n)}) \\
&= nN(r, \psi) + N\left(r, \frac{1}{\psi^{(n+1)}}\right) + O(\log r) \\
&= nT(r, \psi) + T\left(r, \frac{1}{\psi^{(n+1)}}\right) + O(\log r) \\
&= (2n+2)T(r, \psi) + O(\log r).
\end{aligned}$$

□

4. Algebraic differential equations of the second order

In 1887, Émile Picard raised the question of whether differential equations of the form

$$w'' = R(z, w, w'), \tag{4.1}$$

where R is rational in w and w' with entire coefficients, have fixed critical points [25]. Paul Painlevé solved this problem in a series of articles published a few years later, see e.g. [24]. By using a rather technical "alpha–method", he finally ended up with 50 possible types of equations. Most of these equations were reducible to a linear differential equation, or were solvable by elliptic functions; it appeared that only six equations produced essentially new classes of functions. The first three equations were discovered by Painlevé himself, but the last three were added later on by R. Gambier. These equations, which are now called the six Painlevé differential equations, are presently under careful study in the theory of second order differential equations. This is the reason why most of this section is devoted to the value distribution of the solutions of the Painlevé differential equations.

4.1. Linear homogeneous case

Before going into the Painlevé equations, we take a brief look at linear homogeneous second order differential equations. More precisely, we consider the differential equation

$$w'' + P(z)w = 0, \tag{4.2}$$

where $P(z) = a_n z^n + \cdots + a_0$ is a polynomial of degree n with $a_n \neq 0$. We show that solutions of (4.2) do not have any small deficient functions, except possibly the constant function zero. For this we need the following theorem by Gundersen, see [9].

Theorem 4.1.1 (Gundersen). *Let $\phi \not\equiv 0$ be a solution of* (4.2)*. Then as $r \to \infty$,*

$$\log M(r, \phi) = (1 + o(1)) \frac{\sqrt{|a_n|}}{\alpha} r^{\alpha},$$

where $\alpha = (n+2)/2$.

Now we are ready to prove our main result of this section.

Theorem 4.1.2. *Let $\phi \not\equiv 0$ be a solution of* (4.2)*, and let $g \in \mathcal{L}_\phi \setminus \{0\}$. Then $\delta(g, \phi) = 0$.*

Proof. Suppose conversely that there exist $g \in \mathcal{L}_\phi \setminus \{0\}$, such that $\delta(g, \phi) > 0$. Then by Theorem 2.1.1, g must also be a solution of (4.2). Clearly (4.2) does not have any nonzero polynomial solutions, so g must be transcendental. Furthermore, by Lemma 1.2.1,

$$\lim_{r\to\infty} \frac{T(r,g)}{T(2r,\phi)} = 0. \tag{4.3}$$

Since clearly both g and ϕ are entire and transcendental, we have by Theorem 1.1.1, and by Theorem 4.1.1 with $\alpha = (n+2)/2$,

$$\begin{aligned}
\lim_{r\to\infty} \frac{T(r,g)}{T(2r,\phi)} &\geq \lim_{r\to\infty} \frac{\frac{1}{3}\log M(\frac{r}{2},g)}{\log M(2r,\phi)} \\
&= \lim_{r\to\infty} \frac{\frac{1}{3}(1+o(1))\frac{\sqrt{|a_n|}}{\alpha}\left(\frac{r}{2}\right)^\alpha}{(1+o(1))\frac{\sqrt{|a_n|}}{\alpha}(2r)^\alpha} \\
&= \lim_{r\to\infty} \frac{1}{3}(1+o(1))2^{-2\alpha} \\
&= \frac{1}{3}2^{-2\alpha} > 0.
\end{aligned}$$

This contradicts equation (4.3), and so $\delta(g,\phi) = 0$ for all $g \in \mathcal{L}_\phi \setminus \{0\}$. □

Remarks. 1. The above result is sharp in the sense that no upper or lower bound for $\delta(0,\phi)$ can be obtained. Namely, $\phi_a(z) := e^z$ is a solution of

$$f'' - f = 0,$$

and $\phi_b(z) := \sin z$ is a solution of

$$f'' + f = 0.$$

Clearly $\delta(0,\phi_a) = 1$, and $\delta(0,\phi_b) = 0$. It is also known [9, p. 288] that the classical Airy differential equation

$$f'' - zf = 0$$

possesses a solution ϕ_c such that $\delta(0,\phi_c) = 1/2$.

2. Since by Theorem 2.2.3 $\Delta(g,\phi) = 0$ for all $g \in \mathbb{C}(z)$ such that g is not a solution of (4.2), and since the only rational solution of (4.2) is the zero solution, $\Delta(g,\phi) = 0$ for all $g \in \mathbb{C}(z)\setminus\{0\}$.

4.2. The first, the second and the fourth Painlevé transcendents

We start our considerations of the Painlevé transcendents by examining the following three differential equations:

$$w'' = 6w^2 + z, \tag{4.4}$$

$$w'' = 2w^3 + zw + \alpha_2, \tag{4.5}$$

$$ww'' = \frac{1}{2}(w')^2 + \frac{3}{2}w^4 + 4zw^3 + 2(z^2 - \alpha_4)w^2 + \beta_4, \tag{4.6}$$

where $\alpha_j, \beta_j \in \mathbb{C}$. These equations, which are called usually as the first, the second, and the fourth Painlevé differential equation, see e.g. [7, p. 157], have been recently under careful study: A. Hinkkanen and I. Laine proved that all solutions of (4.4) and (4.5) are meromorphic [17], and, later on by using different technique, N. Steinmetz showed that all solutions of (4.6) are meromorphic [34]. He also proved that all solutions of (4.4), (4.5) and (4.6) are of finite order [35]. Also S. Shimomura showed this fact by using different method in [27]. Furthermore, Shimomura proved a number of results about the value distribution of the solutions of equations (4.4), (4.5) and (4.6), see, for instance, [28], [29] or [31].

Despite all of this research activity, the value distribution of the derivatives of the Painlevé transcendents has remained, so far, almost unnoticed. This motivates us to focus our attention in this section to this topic. For the basic properties of meromorphic derivatives see, for instance, [37, pp. 121–131].

Let ϕ_1, ϕ_2 and ϕ_4 denote arbitrary transcendental meromorphic solutions of (4.4), (4.5) and (4.6), respectively. By using Theorem 2.2.3, we are able to prove the following theorem about the Valiron deficiencies of ϕ_1, ϕ_2 and ϕ_4. Shimomura has proved essentially the same result in [28], for the usual deficiency δ.

Theorem 4.2.1.

1. *$\Delta(g, \phi_1) = 0$ for every rational function g.*
2. *If $\alpha_2 \notin \mathbb{Z}$, then $\Delta(g, \phi_2) = 0$ for every rational function g.*
3. *If $\alpha_2 \in \mathbb{Z}$, then $\Delta(g, \phi_2) = 0$ for every rational function g, other than the unique rational solution of* (4.5).
4. *$\Delta(g, \phi_4) = 0$ for every rational function g, which is not a solution of* (4.6).
5. *$\Delta(\infty, \phi_1) = \Delta(\infty, \phi_2) = \Delta(\infty, \phi_4) = 0$.*

Proof. We know that (4.4) has no rational solutions. Also, by [23], (4.5) has a rational solution if, and only if, $\alpha \in \mathbb{Z}$, and then the rational solution is unique. Hence, the four first assertions follow by Theorem 2.2.3. The fifth assertion follows by Lemma 2.2.1. □

The following lemma is fundamental for our considerations of the value distribution of derivatives of the Painlevé transcendents. To build up a solid basis for subsequent results, the proof of this lemma is written out in detail, even though it is rather long and technical.

Lemma 4.2.2. *Let P be a polynomial, let $j = 1, 2, 4$, and let $n \in \mathbb{N} \cup \{0\}$. Then*

$$m(r, \phi_j^{(n)}) = O(\log r), \tag{4.7}$$

and

$$m\left(r, \frac{1}{\phi_j^{(n)} - P}\right) = O(\log r), \tag{4.8}$$

provided that $\alpha_2, \beta_4 \neq 0$ in the cases $j = 2, 4$, respectively.

Proof. First observe that the case $n = 0$ follows directly by Lemma 2.2.1 for (4.7). Hence, by the lemma of the logarithmic derivative,

$$m(r, \phi_j^{(n)}) \leq m\left(r, \frac{\phi_j^{(n)}}{\phi_j}\right) + m(r, \phi_j) = O(\log r)$$

for all $j = 1, 2, 4$, and for all $n \in \mathbb{N} \cup \{0\}$.

Consider next the assertion (4.8) for the case $j = 1$ and $P \equiv 0$. We have,

$$m\left(r, \frac{1}{\phi_1}\right) = O(\log r)$$

by Lemma 2.2.2. By differentiating (4.4), and dividing by ϕ_1', we have

$$\frac{1}{\phi_1'} = \frac{\phi_1'''}{\phi_1'} - 12\phi_1.$$

Hence,

$$m\left(r, \frac{1}{\phi_1'}\right) \leq m\left(r, \frac{\phi_1'''}{\phi_1'}\right) + m(r, \phi_1) + O(1) = O(\log r)$$

by the lemma of the logarithmic derivative, and by (4.7).

Assume now that we have proved

$$m\left(r, \frac{1}{\phi_1^{(n)}}\right) = O(\log r) \tag{4.9}$$

for all $n = 1, \ldots, k$. By differentiating equation (4.4) $2k$ times, we obtain

$$\phi_1^{(2k+2)} = 6 \sum_{j=0}^{2k} \binom{2k}{j} \phi_1^{(j)} \phi_1^{(2k-j)},$$

so that

$$\frac{\left(\phi_1^{(k)}\right)^2}{\phi_1^{(k+1)}} = \frac{1}{\binom{2k}{k}} \left(\frac{1}{6} \frac{\phi_1^{(2k+2)}}{\phi_1^{(k+1)}} - \sum_{\substack{j=0 \\ j \neq k}}^{2k} \binom{2k}{j} \frac{\phi_1^{(j)} \phi_1^{(2k-j)}}{\phi_1^{(k+1)}} \right). \tag{4.10}$$

Hence, by (4.7), (4.9), (4.10) and by the lemma of the logarithmic derivative,

$$\begin{aligned} m\left(r, \frac{1}{\phi_1^{(k+1)}}\right) &\leq m\left(r, \frac{(\phi_1^{(k)})^2}{\phi_1^{(k+1)}}\right) + m\left(r, \frac{1}{(\phi_1^{(k)})^2}\right) \\ &= m\left(r, \frac{(\phi_1^{(k)})^2}{\phi_1^{(k+1)}}\right) + O(\log r) = O(\log r). \end{aligned}$$

Thus

$$m\left(r, \frac{1}{\phi_1^{(n)}}\right) = O(\log r) \tag{4.11}$$

for all $n \in \mathbb{N} \cup \{0\}$ by the induction principle.

Consider now the case $j = 2$ and $P \equiv 0$. We have,

$$m\left(r, \frac{1}{\phi_2}\right) = O(\log r) \tag{4.12}$$

by Lemma 2.2.2, since we assumed $\alpha_2 \neq 0$. By differentiating (4.5), we have

$$\phi_2''' = \phi_2 + z\phi_2' + 6\phi_2^2\phi_2'.$$

By dividing this with ϕ_2', we obtain

$$\frac{\phi_2}{\phi_2'} = \frac{\phi_2'''}{\phi_2'} - z - 6\phi_2^2. \tag{4.13}$$

Hence, by (4.12) and (4.13),

$$m\left(r, \frac{1}{\phi_2'}\right) \leq m\left(r, \frac{\phi_2}{\phi_2'}\right) + m\left(r, \frac{1}{\phi_2}\right) = O(\log r).$$

Similarly as before, we assume that

$$m\left(r, \frac{1}{\phi_2^{(n)}}\right) = O(\log r) \tag{4.14}$$

has been proved for all $n = 1, \ldots, k$. By differentiating equation (4.5) $3k$ times, we obtain

$$\phi_2^{(3k+2)} = z\phi_2^{(3k)} + 3k\phi_2^{(3k-1)} + 2\sum_{j=0}^{3k}\sum_{m=0}^{j}\binom{3k}{j}\binom{j}{m}\phi_2^{(3k-j)}\phi_2^{(m)}\phi_2^{(j-m)},$$

and so

$$\begin{aligned} \frac{(\phi_2^{(k)})^3}{\phi_2^{(k+1)}} = \frac{1}{\binom{3k}{2k}\binom{2k}{k}}\Bigg(&\frac{1}{2}\frac{\phi_2^{(3k+2)}}{\phi_2^{(k+1)}} - \frac{1}{2}\frac{z\phi_2^{(3k)}}{\phi_2^{(k+1)}} - \frac{1}{2}\frac{3k\phi_2^{(3k-1)}}{\phi_2^{(k+1)}} \\ &- \sum_{j=0}^{3k}\sum_{\substack{m=0 \\ (m,j)\neq(k,2k)}}^{j}\binom{3k}{j}\binom{j}{m}\frac{\phi_2^{(3k-j)}\phi_2^{(m)}\phi_2^{(j-m)}}{\phi_2^{(k+1)}}\Bigg). \end{aligned} \tag{4.15}$$

Every term in the numerator of the last term on the right hand side of (4.15) has at least one function $\phi_2^{(l)}$ as a factor, such that $l \geq k+1$. Hence by (4.7), (4.14), and by the lemma of the logarithmic derivative,

$$m\left(r, \frac{1}{\phi_2^{(k+1)}}\right) \leq m\left(r, \frac{(\phi_2^{(k)})^3}{\phi_2^{(k+1)}}\right) + m\left(r, \frac{1}{(\phi_2^{(k)})^3}\right)$$
$$= m\left(r, \frac{(\phi_2^{(k)})^3}{\phi_2^{(k+1)}}\right) + O(\log r) = O(\log r).$$

Thus,

$$m\left(r, \frac{1}{\phi_2^{(n)}}\right) = O(\log r) \tag{4.16}$$

for all $n \in \mathbb{N} \cup \{0\}$ by the induction principle.

Consider now the case $j = 4$ and $P \equiv 0$. We have,

$$m\left(r, \frac{1}{\phi_4}\right) = O(\log r) \tag{4.17}$$

by Lemma 2.2.2, since we assumed $\beta_4 \neq 0$. By differentiating (4.6) and dividing by ϕ_4, we have

$$\phi_4''' = 6\phi_4'\phi_4^2 + 4\phi_4^2 + 12z\phi_4'\phi_4 + 4z\phi_4 + 4z^2\phi_4' - 4\alpha_4\phi_4'. \tag{4.18}$$

By differentiating this again, and dividing by $4\phi_4'$, we have

$$\frac{\phi_4}{\phi_4'} = \frac{1}{4}\frac{\phi_4^{(4)}}{\phi_4'} - \frac{3}{2}\frac{\phi_4''}{\phi_4'}\phi_4^2 - 3\phi_4'\phi_4 - 5\phi_4 - 3z\frac{\phi_4''}{\phi_4'}\phi_4 - 3z\phi_4' - 3z - (z^2 - \alpha_4)\frac{\phi_4''}{\phi_4'},$$

and so we have again, similarly as before, by (4.17) and by the lemma of the logarithmic derivative,

$$m\left(r, \frac{1}{\phi_4'}\right) = O(\log r).$$

Assume now that

$$m\left(r, \frac{1}{\phi_4^{(n)}}\right) = O(\log r) \tag{4.19}$$

has been proved for all $n = 1, \ldots, k$. Differentiating (4.18) $3k - 1$ times, we have

$$\begin{aligned}
\phi_4^{(3k+2)} &= 6\sum_{j=0}^{3k-1}\binom{3k-1}{j}(\phi_4')^{(3k-j-1)}(\phi_4^2)^{(j)} + 4\sum_{j=0}^{3k-1}\binom{3k-1}{j}\phi_4^{(j)}\phi_4^{(3k-j-1)} \\
&+ 12\sum_{j=0}^{3k-1}\binom{3k-1}{j}(z\phi_4')^{(3k-j-1)}\phi_4^{(j)} + 4z\phi_4^{(3k-1)} + (12k-4)\phi_4^{(3k-2)} \\
&+ 4\sum_{j=0}^{3k-1}\binom{3k-1}{j}(z^2)^{(3k-j-1)}(\phi_4')^{(j)} - 4\alpha_4\phi_4^{(3k)} \\
&= 6\sum_{j=0}^{3k-1}\sum_{m=0}^{j}\binom{3k-1}{j}\binom{j}{m}\phi_4^{(3k-j)}\phi_4^{(m)}\phi_4^{(j-m)} + 4\sum_{j=0}^{3k-1}\binom{3k-1}{j}\phi_4^{(j)}\phi_4^{(3k-j-1)} \\
&+ 12\sum_{j=0}^{3k-1}\binom{3k-1}{j}z\phi_4^{(3k-j)}\phi_4^{(j)} + 12\sum_{j=0}^{3k-1}\binom{3k-1}{j}(3k-j-1)\phi_4^{(3k-j-1)}\phi_4^{(j)} \\
&+ 4z\phi_4^{(3k-1)} + (12k-4)\phi_4^{(3k-2)} + 4\sum_{j=0}^{3k-1}\binom{3k-1}{j}(z^2)^{(3k-j-1)}(\phi_4)^{(j+1)} \\
&- 4\alpha_4\phi_4^{(3k)},
\end{aligned}$$

and so

$$6\binom{3k-1}{2k}\binom{2k}{k}\frac{(\phi_4^{(k)})^3}{\phi_4^{(k+1)}} = \frac{\phi_4^{(3k+2)}}{\phi_4^{(k+1)}} - \frac{Q(z,\phi_4)}{\phi_4^{(k+1)}}, \tag{4.20}$$

where $Q(z,\phi_4)$ is a differential polynomial in ϕ_4 and its derivatives with polynomial coefficients. Moreover, every term of Q has some $\phi_4^{(l)}$ as a factor, where $l \geq k+1$. Hence by (4.7), (4.19), (4.20) and by the lemma of the logarithmic derivative,

$$\begin{aligned}
m\left(r, \frac{1}{\phi_4^{(k+1)}}\right) &\leq m\left(r, \frac{(\phi_4^{(k)})^3}{\phi_4^{(k+1)}}\right) + m\left(r, \frac{1}{(\phi_4^{(k)})^3}\right) \\
&= m\left(r, \frac{(\phi_4^{(k)})^3}{\phi_4^{(k+1)}}\right) + O(\log r) = O(\log r).
\end{aligned}$$

Therefore

$$m\left(r, \frac{1}{\phi_4^{(n)}}\right) = O(\log r) \tag{4.21}$$

for all $n \in \mathbb{N} \cup \{0\}$ by the induction principle.

Suppose now that $P \not\equiv 0$, and that $\deg(P) = p$. Then we have for all

$n \in \mathbb{N} \cup \{0\}$, and for all $j = 1, 2, 4$,

$$m\left(r, \frac{1}{\phi_j^{(n)} - P}\right) \leq m\left(r, \frac{(\phi_j^{(n)} - P)^{(p+1)}}{\phi_j^{(n)} - P}\right) + m\left(r, \frac{1}{\phi_j^{(n+p+1)}}\right) = O(\log r),$$

by (4.11), (4.16) and (4.21) and by the lemma of the logarithmic derivative. □

The following theorem, similar to Theorem 4.2.1, is an immediate consequence of Lemma 4.2.2.

Theorem 4.2.3. *Let $n \in \mathbb{N} \cup \{0\}$. Then*

1. $\Delta(P, \phi_1^{(n)}) = 0$ *for every polynomial* P.
2. *If* $\alpha_2 \neq 0$, *then* $\Delta(P, \phi_2^{(n)}) = 0$ *for every polynomial* P.
3. *If* $\beta_4 \neq 0$, *then* $\Delta(P, \phi_4^{(n)}) = 0$ *for every polynomial* P.
4. $\Delta(\infty, \phi_1^{(n)}) = \Delta(\infty, \phi_2^{(n)}) = \Delta(\infty, \phi_4^{(n)}) = 0.$

Note that in Theorem 4.2.3 we do not need to assume that P does not satisfy the corresponding equation. Hence also the case $n = 0$ gives interesting new information compared to Theorem 4.2.1, as the following example demonstrates.

Example 4.2.4. Consider the following two special cases of the fourth Painlevé differential equation: The equation

$$ww'' = \frac{1}{2}(w')^2 + \frac{3}{2}w^4 + 4zw^3 + 2z^2w^2 - 2 \tag{4.22}$$

has a polynomial solution $P_1(z) := -2z$, and

$$ww'' = \frac{1}{2}(w')^2 + \frac{3}{2}w^4 + 4zw^3 + 2z^2w^2 - \frac{2}{9}, \tag{4.23}$$

has a polynomial solution $P_2(z) := -\frac{2}{3}z$. Let ϕ_{4a} and ϕ_{4b} denote transcendental solutions of (4.22) and (4.23), respectively. Then Theorem 4.2.1 does not tell anything about $\Delta(P_1, \phi_{4a})$, or $\Delta(P_2, \phi_{4b})$, but by Theorem 4.2.3 we know that, in fact, $\Delta(P_1, \phi_{4a}) = \Delta(P_2, \phi_{4b}) = 0$.

We now use Lemma 4.2.2 to investigate the value distribution of $\phi_j^{(n)}$, $j = 1, 2, 4$. Note that since the proofs of the following theorems are very similar to those in Section 3.3, we omit most of the details.

Theorem 4.2.5. *Let $n \in \mathbb{N} \cup \{0\}$. Then*

$$\Phi_{(n)}(\phi_1, P) = \frac{n+2}{2}$$

for all polynomials P. If $\alpha_2, \beta_4 \neq 0$, then

$$\Phi_{(n)}(\phi_j, P) = n+1,$$

where $j = 2, 4$.

Proof. Since all poles of ϕ_1 are double,

$$N(r, \phi_1^{(n)}) = \frac{n+2}{2} T(r, \phi_1) + O(\log r).$$

Hence, by Lemma 4.2.2,

$$N\left(r, \frac{1}{\phi_1^{(n)} - P}\right) = \frac{n+2}{2} T(r, \phi_1) + O(\log r),$$

and so $\Phi_{(n)}(\phi_1, P) = \frac{n+2}{2}$.

Similarly, since we know that all poles of ϕ_2 are simple,

$$N(r, \phi_2^{(n)}) = (n+1)T(r, \phi_2) + O(\log r),$$

and so by Lemma 4.2.2 again,

$$N\left(r, \frac{1}{\phi_2^{(n)} - P}\right) = (n+1)T(r, \phi_2) + O(\log r).$$

Hence, $\Phi_{(n)}(\phi_2) = n+1$.

The assertion for ϕ_4 follows with completely similar reasoning by observing that all poles of ϕ_4 are also simple. □

Remarks. 1.Theorem 4.2.5 reduces in the case $n = 1$, $j = 1$ and $P \equiv 0$ to [21, Theorem 9.2.2].

2. Since $\Phi_{(n)}(f, P) \leq n+1$ for all transcendental meromorphic functions f by definition, we can see that $\Phi_{(n)}(\phi_j, P)$ is, in this sense, maximal, whenever $j = 2, 4$ and $\alpha_2, \beta_4 \neq 0$.

Similarly, we may consider the growth of the functions $\phi_j^{(n)}$, $j = 1, 2, 4$.

Theorem 4.2.6. *Let $n \in \mathbb{N}$, and let the assumptions of Theorem 4.2.5 hold. Then*

$$T(r, \phi_1^{(n)}) = \frac{n+2}{2} T(r, \phi_1) + O(\log r),$$

and

$$T(r, \phi_j^{(n)}) = (n+1)T(r, \phi_j) + O(\log r)$$

for $j = 2, 4$.

As we can see from Theorems 4.2.3 and 4.2.5, the value distribution of the derivatives of the first, the second and the fourth Painlevé transcendents is extremely regular. The following theorem demonstrates this regularity even further.

Theorem 4.2.7. *Let $\alpha_2, \beta_4 \neq 0$, and let $n \in \mathbb{N} \cup \{0\}$. Then*

$$\Phi_e(\phi_1^{(n)}) = \frac{n+3}{n+2}$$

and

$$\Phi_e(\phi_j^{(n)}) = \frac{n+2}{n+1}$$

for $j = 2, 4$.

Proof. Consider the case $j = 1$. By Lemma 4.2.2 and Theorem 4.2.6, we have

$$\Phi_e(\phi_1^{(n)}) = \liminf_{r\to\infty} \frac{N\left(r, \frac{1}{(\phi_1^{(n)})'}\right)}{T(r, \phi_1^{(n)})} = \liminf_{r\to\infty} \frac{\frac{n+3}{2}T(r,\phi_1) + O(\log r)}{\frac{n+2}{2}T(r,\phi_1) + O(\log r)} = \frac{n+3}{n+2}.$$

The proofs for the other cases are completely similar, and hence they are omitted here. □

Now we further investigate the value distribution of the derivatives of the solutions of the Painlevé equations by proving the following version of the second main theorem for $\phi_j^{(n)}$, where $j = 1, 2, 4$ and $n \in \mathbb{N} \cup \{0\}$.

Theorem 4.2.8. *Let $P_1, \ldots, P_q$ be distinct complex polynomials, and let $\alpha_2, \beta_4 \neq 0$. Then for all $n \in \mathbb{N} \cup \{0\}$,*

$$\sum_{k=1}^{q} m\left(r, \frac{1}{\phi_1^{(n)} - P_k}\right) + N_B(r, \phi_1^{(n)}) = (n+2)T(r, \phi_1) + O(\log r),$$

and

$$\sum_{k=1}^{q} m\left(r, \frac{1}{\phi_j^{(n)} - P_k}\right) + N_B(r, \phi_j^{(n)}) = (2n+2)T(r, \phi_j) + O(\log r),$$

where $j = 2, 4$.

Proof. Consider first the case $j = 1$. Note that since all poles of ϕ_1 are double, we have $N(r, \phi_1^{(n)}) - \overline{N}(r, \phi_1^{(n)}) = \frac{n+1}{2}N(r, \phi_1)$. Now we have by Lemma 4.2.2

and Theorem 4.2.6, that

$$\begin{aligned}\sum_{k=1}^{q} m\left(r, \frac{1}{\phi_1^{(n)} - P_k}\right) + N_B(r, \phi_1^{(n)}) &= \sum_{k=1}^{q} m\left(r, \frac{1}{\phi_1^{(n)} - P_k}\right) + N\left(r, \frac{1}{(\phi_1^{(n)})'}\right) \\ &\quad + N(r, \phi_1^{(n)}) - \overline{N}(r, \phi_1^{(n)}) \\ &= N\left(r, \frac{1}{\phi_1^{(n+1)}}\right) + \frac{n+1}{2} N(r, \phi_1) + O(\log r) \\ &= \frac{n+3}{2} T(r, \phi_1) + \frac{n+1}{2} T(r, \phi_1) + O(\log r) \\ &= (n+2) T(r, \phi_1) + O(\log r).\end{aligned}$$

The cases $j = 2, 4$ can be proved similarly. □

4.3. The third, and the fifth Painlevé transcendents

We continue to consider the derivatives of solutions of Painlevé differential equations. In this section, the following two types of equations are investigated:

$$w'' = \frac{(w')^2}{w} - \frac{w'}{z} + \frac{\alpha_3 w^2 + \beta_3}{z} + \gamma_3 w^3 + \frac{\delta_3}{w}, \tag{4.24}$$

$$\begin{aligned} w'' = {} & \left(\frac{1}{2w} + \frac{1}{w-1}\right)(w')^2 - \frac{w'}{z} + \frac{(w-1)^2}{z^2 w}(\alpha_5 w^2 + \beta_5) \\ & + \frac{\gamma_5 w}{z} + \frac{\delta_5 w(w+1)}{w-1}, \end{aligned} \tag{4.25}$$

where $\alpha_j, \beta_j, \gamma_j, \delta_j \in \mathbb{C}$. They are usually called as the third and the fifth Painlevé differential equation, see e.g. [7, p. 157].

Unlike the cases of the first, the second, and the fourth Painlevé differential equations, it is immediately seen that not all solutions of (4.24) are meromorphic. As an example, take $w(z) = Cz^{1/3}$ with $C^3 = -\delta_3/\alpha_3$ in the case $\beta_3 = \gamma_3 = 0$, and $\alpha_3\delta_3 \neq 0$. However, by making a transformation

$$W(\zeta) = zw(z) = e^{\zeta/2} w(e^{\zeta/2}), \quad z = e^{\zeta/2},$$

and by replacing $(\alpha_3, \beta_3, \gamma_3, \delta_3)$ with $(4\alpha_3, 4\beta_3, 4\gamma_3, 4\delta_3)$, we have an equation of the form

$$ww'' = (w')^2 + \alpha_3 w^3 + \gamma_3 w^4 + \beta_3 e^z w + \delta_3 e^{2z}. \tag{4.26}$$

Hinkkanen and Laine proved in [16] that all solutions of this modified third Painlevé differential equation are meromorphic.

Similarly to the third Painlevé equation, not all solutions of (4.25) are meromorphic, see for example [15]. But by making a transformation

$$W(\zeta) = w(z) = w(e^{\zeta}), \quad z = e^{\zeta},$$

we have an equation of the form

$$w'' = \left(\frac{1}{2w} + \frac{1}{w-1}\right)(w')^2 + (w-1)^2\left(\alpha_5 w + \frac{\beta_5}{w}\right) \\ + \gamma_5 e^z w + \frac{\delta_5 e^{2z} w(w+1)}{w-1}. \tag{4.27}$$

All solutions of this modified fifth Painlevé equation are meromorphic by [15].

Let now ϕ_3 and ϕ_5 denote arbitrary transcendental meromorphic solutions of (4.26) and (4.27), respectively. Shimomura proved [26] the following result about the growth of ϕ_3 and ϕ_5.

Theorem 4.3.1 (Shimomura). *For some $\Gamma > 0$,*

$$T(r, \phi_j) = O(e^{\Gamma r}),$$

where $j = 3, 5$, and Γ is independent of $\phi_j(z)$.

Theorem 4.3.1 has, as was pointed out in [26], the following corollary.

Corollary 4.3.2 (Shimomura).

a) If $(\alpha_3, \gamma_3) \neq (0,0)$, then $m(r, \phi_3) = O(r)$.

b) If $\delta_3 \neq 0$, then $m\left(r, \frac{1}{\phi_3}\right) = O(r)$.

c) If $\alpha_5 \neq 0$, then $m(r, \phi_5) = O(r)$.

d) If $\beta_5 \neq 0$, then $m\left(r, \frac{1}{\phi_5}\right) = O(r)$.

We also have by the lemma of the logarithmic derivative,

$$m\left(r, \frac{f^{(n)}}{f}\right) = O(r), \tag{4.28}$$

where $n \in \mathbb{N}$, and $f = \phi_j + g$, $j = 3, 5$, for some rational function g.

From now on, we assume that ϕ_3 and ϕ_5 are admissible solutions of (4.26) and (4.27), respectively, in the sense that

$$\frac{r}{T(r, \phi_j)} \longrightarrow 0$$

as $r \to \infty$, outside of a possible exceptional set of finite linear measure. Denote $s := 2r$. Then by Lemma 1.2.1

$$\frac{1}{2}\frac{s}{T(s, \phi_j)} \longrightarrow 0$$

as $s \to \infty$ without an exceptional set. Hence, from now on, we may omit the exceptional set when we say that ϕ_j, $j = 3, 5$, is an admissible solution of either (4.26), or (4.27).

The following theorem, which is analogous to Theorem 4.2.1, considers the Valiron deficiency of the solutions of equations (4.26) and (4.27).

Theorem 4.3.3.

1. *If $(\beta_3, \delta_3) \neq (0,0)$, then $\Delta(g, \phi_3) = 0$ for every rational function g.*
2. *If $\beta_5\delta_5 \neq 0$ and either $\gamma_5 \neq 0$, or $\alpha_5 \neq -\beta_5$, then $\Delta(g, \phi_5) = 0$ for every rational function g.*
3. *If $(\alpha_3, \gamma_3) \neq (0,0)$, then $\Delta(\infty, \phi_3) = 0$.*
4. *If $\alpha_5 \neq 0$, then $\Delta(\infty, \phi_5) = 0$.*

Proof. Let g be a rational function. We have, by substituting $w = F_3 + g$ into (4.26),

$$Q(z, F_3, F_3', F_3'') + gg'' - (g')^2 - \alpha_3 g^3 - \gamma_3 g^4 - \beta_3 e^z g - \delta_3 e^{2z} = 0, \qquad (4.29)$$

where Q is a differential polynomial in F_3 and its derivatives with coefficients in the field $\{h : T(r,h) = O(r)\}$. Furthermore, $Q(z,0,0,0) \equiv 0$, and $F_3 = \phi_3 - g$ is a solution of (4.29). Since

$$gg'' - (g')^2 - \alpha_3 g^3 - \gamma_3 g^4 - \beta_3 e^z g - \delta_3 e^{2z} \not\equiv 0,$$

whenever $(\beta_3, \delta_3) \neq (0,0)$ and $g \not\equiv 0$ is rational, we have by taking (4.28) into account in the proof of Lemma 1.3.2, that

$$m\left(r, \frac{1}{\phi_3 - g}\right) = m\left(r, \frac{1}{F_3}\right) = O(r).$$

Hence $\Delta(g, \phi_3) = 0$, whenever $g \not\equiv 0$. Assume now that $g \equiv 0$. If $\delta_3 \neq 0$, the assertion follows by Corollary 4.3.2. Hence suppose that $\beta_3 \neq 0$. By the transformation $w = e^z/W$ [32, Lemma 3.1], equation (4.26) is changed into

$$W'' = \frac{(W')^2}{W} - \beta_3 W^2 - \delta_3 W^3 - \alpha_3 e^z - \frac{\gamma_3 e^{2z}}{W}.$$

Therefore, by the Clunie lemma,

$$m\left(r, \frac{1}{\phi_3}\right) = m\left(r, \frac{W}{e^z}\right) = m(r, W) + O(r) = O(r).$$

This completes the proof of the first assertion.

The final three assertions are obtained by a similar reasoning as in the proof of the first assertion, and hence we omit the details here. □

The following lemma is similar to Lemma 4.2.2 for the solutions of the modified third and fifth Painlevé equations.

Lemma 4.3.4. *Let P be a polynomial, let $j = 3, 5$, and let $n \in \mathbb{N} \cup \{0\}$. Then*

$$m(r, \phi_j^{(n)}) = O(r), \tag{4.30}$$

when $(\alpha_3, \gamma_3) \neq (0,0)$ and $\alpha_5 \neq 0$, and

$$m\left(r, \frac{1}{\phi_j^{(n)} - P}\right) = O(r). \tag{4.31}$$

whenever $(\beta_3, \delta_3) \neq (0,0)$ and $(\alpha_3, \gamma_3) \neq (0,0)$ in the case $j = 3$, and $\alpha_5\beta_5\delta_5 \neq 0$ and either $\gamma_5 \neq 0$, or $\alpha_5 \neq -\beta_5$, in the case $j = 5$.

Remark. Note that Shimomura already proved the assertion for ϕ_3 and ϕ_5 in the case $n = 1$ and $P \equiv 0$, see [32] and [30], respectively. We still include the proof of these cases here for completeness.

Proof. Concerning (4.30), the case $n = 0$ is contained in Corollary 4.3.2. Hence, by (4.28),

$$m(r, \phi_j^{(n)}) \leq m\left(r, \frac{\phi_j^{(n)}}{\phi_j}\right) + m(r, \phi_j) = O(r)$$

for $j = 3, 5$.

Consider now (4.31) for $j = 3$, and assume first that $P \equiv 0$. If $\delta_3 \neq 0$,

$$m\left(r, \frac{1}{\phi_3}\right) = O(r) \tag{4.32}$$

by Corollary 4.3.2. If $\delta_3 = 0$, then $\beta_3 \neq 0$, by assumption, and we obtain (4.32) by repeating the reasoning in the end of the proof of Theorem 4.3.3. By differentiating (4.26), we have

$$\phi_3'''\phi_3 - \phi_3''\phi_3' - 3\alpha_3\phi_3^2\phi_3' - 4\gamma_3\phi_3^3\phi_3' - \beta_3 e^z\phi_3 - \beta_3 e^z\phi_3' - 2\delta_3 e^{2z} = 0. \tag{4.33}$$

Differentiating (4.33), we get

$$\begin{aligned} 4\delta_3 e^{2z} = {} & \phi_3^{(4)}\phi_3 - (\phi_3'')^2 - 6\alpha_3\phi_3(\phi_3')^2 - 3\alpha_3\phi_3^2\phi_3'' - 12\gamma_3\phi_3^2(\phi_3')^2 \\ & - 4\gamma_3\phi_3^3\phi_3'' - \beta_3 e^z\phi_3 - 2\beta_3 e^z\phi_3' - \beta_3 e^z\phi_3''. \end{aligned} \tag{4.34}$$

By subtracting (4.33) from (4.34), we have

$$\begin{aligned} 2\delta_3 e^{2z} = {} & \phi_3^{(4)}\phi_3 - \phi_3'''\phi_3 + \phi_3''\phi_3' - (\phi_3'')^2 - 4\gamma_3\phi_3^3\phi_3'' - 3\alpha_3\phi_3^2\phi_3'' \\ & - \beta_3 e^z\phi_3'' - \beta_3 e^z\phi_3' + 3\alpha_3\phi_3^2\phi_3' + 4\gamma_3\phi_3^3\phi_3' - 6\alpha_3\phi_3(\phi_3')^2 - 12\gamma_3\phi_3^2(\phi_3')^2, \end{aligned}$$

and so

$$m\left(r, \frac{1}{\phi_3'}\right) = m\left(r, \frac{2\delta_3 e^{2z}}{\phi_3'}\right) + O(r) = O(r),$$

by Corollary 4.3.2 and (4.28).

Assume now that

$$m\left(r, \frac{1}{\phi_3^{(n)}}\right) = O(r) \tag{4.35}$$

has been proved for all $n = 1, \ldots, k$. By differentiating (4.26) $4k+1+m$ times, where $m = 0, \ldots, k+2$, we obtain the following system of $k+3$ equations:

$$2^{4k+1+m}\delta_3 e^{2z} = -\beta_3 e^z \sum_{j=0}^{k} \binom{4k+1+m}{j} \phi_3^{(j)} + Q_m(z, \phi_3), \tag{4.36}$$

where Q_m is a differential polynomial in ϕ_3 and its derivatives with coefficients in the field $\{h : T(r,h) = O(r)\}$, such that each term of Q_m has at least one $\phi_3^{(l)}$, $l \geq k+1$, as a factor. Now choose $a_m \in \mathbb{R}$, where $m = 0, \ldots, k+2$, such that the following system of linear equations is satisfied:

$$\begin{cases} \binom{4k+1}{0}a_0 + \binom{4k+2}{0}a_1 + \cdots + \binom{5k+3}{0}a_{k+2} = 0 \\ \binom{4k+1}{1}a_0 + \binom{4k+2}{1}a_1 + \cdots + \binom{5k+3}{1}a_{k+2} = 0 \\ \vdots \qquad\qquad\qquad\qquad\qquad\qquad \vdots \\ \binom{4k+1}{k}a_0 + \binom{4k+2}{k}a_1 + \cdots + \binom{5k+3}{k}a_{k+2} = 0 \\ 2^{4k+1}a_0 + 2^{4k+2}a_1 + \cdots + 2^{5k+3}a_{k+2} \quad = 1. \end{cases} \tag{4.37}$$

We may choose such a_m's, since all rows of (4.37) are linearly independent, and (4.37) has more indeterminates than equations. Now, by (4.36), we have

$$\begin{aligned} \sum_{m=0}^{k+2} 2^{4k+1+m}\delta_3 a_m e^{2z} &= -\beta_3 e^z \sum_{m=0}^{k+2}\sum_{j=0}^{k} a_m \binom{4k+1+m}{j} \phi_3^{(j)} + \sum_{m=0}^{k+2} a_m Q_m(z,\phi_3) \\ &= -\beta_3 e^z \sum_{j=0}^{k}\sum_{m=0}^{k+2} a_m \binom{4k+1+m}{j} \phi_3^{(j)} + \sum_{m=0}^{k+2} a_m Q_m(z,\phi_3), \end{aligned}$$

and so, by (4.37),

$$\delta_3 e^{2z} = \sum_{m=0}^{k+2} a_m Q_m(z,\phi_3).$$

Hence,

$$m\left(r, \frac{1}{\phi_3^{(k+1)}}\right) = m\left(r, \frac{\delta_3 e^{2z}}{\phi_3^{(k+1)}}\right) + O(r) = O(r),$$

and so

$$m\left(r, \frac{1}{\phi_3^{(n)}}\right) = O(r) \tag{4.38}$$

for all $n \in \mathbb{N} \cup \{0\}$ by the induction principle.

Consider next the case $j = 5$, assuming $P \equiv 0$. Then

$$m\left(r, \frac{1}{\phi_5}\right) = O(r)$$

by Corollary 4.3.2. We also have by (4.27),

$$\begin{aligned}\phi_5\phi_5''(\phi_5 - 1) = (\frac{3}{2}\phi_5 - \frac{1}{2})(\phi_5')^2 + \alpha_5\phi_5^5 - \beta_5 - 3\phi_5(\alpha_5\phi_5^2 + \beta_5)(\phi_5 - 1) \\ - \phi_5^2(\alpha_5 - \beta_5\phi_5) + \gamma_5 e^z\phi_5^2(\phi_5 - 1) + \delta_5 e^{2z}\phi_5^2(\phi_5 + 1).\end{aligned} \tag{4.39}$$

By differentiating (4.39), we obtain

$$2\delta_5 e^{2z}\phi_5^2(\phi_5 + 1) + \gamma_5 e^z\phi_5^2(\phi_5 - 1) = Q(z, \phi_5), \tag{4.40}$$

where Q is a differential polynomial in ϕ_5 and its derivatives with coefficients in the field $\{h : T(r, h) = O(r)\}$, such that each term of Q has at least one $\phi_5^{(l)}$, $l \geq 1$, as a factor. By multiplying (4.40) with e^{-z}, and by differentiating, we have

$$\begin{aligned}2\delta_5 e^z\phi_5^2(\phi_5 + 1) = e^{-z}\frac{d}{dz}Q(z, \phi_5) - e^{-z}Q(z, \phi_5) - \gamma_5\phi_5'\phi_5(3\phi_5 - 2) \\ - 2\delta_5 e^z\phi_5'\phi_5(3\phi_5 + 2).\end{aligned} \tag{4.41}$$

We also have,

$$\begin{aligned}m\left(r, \frac{1}{\phi_5'}\right) \leq m\left(r, \frac{\delta_5 e^z\phi_5^2(\phi_5 + 1)}{\phi_5'}\right) + 2m\left(r, \frac{1}{\phi_5}\right) \\ + m\left(r, \frac{1}{\phi_5 + 1}\right) + O(r).\end{aligned} \tag{4.42}$$

Now, we easily obtain $m(r, 1/(\phi_5 + 1)) = O(r)$ by applying the remark following Lemma 2.2.2, and by remembering our preassumptions on the parameter values. Hence,

$$m\left(r, \frac{1}{\phi_5'}\right) = O(r),$$

by Corollary 4.3.2, and by (4.28), (4.41) and (4.42).

Assume now that

$$m\left(r, \frac{1}{\phi_5^{(n)}}\right) = O(r) \tag{4.43}$$

has been proved for all $n = 1, \dots, k$. Denote

$$F_\delta := \phi_5^2(\phi_5 + 1),$$

and

$$F_\gamma := \phi_5^2(\phi_5 - 1).$$

By differentiating equation (4.39) $5k + m + 1$ times, where $m = 0, \dots, 6k + 3$, we obtain the following system of $6k + 4$ equations:

$$\begin{aligned} Q_m(z, \phi_5) = \delta_5 e^{2z} &\left(\sum_{j=0}^{5k+m+1} 2^{5k+m+1-j} \binom{5k+m+1}{j} F_\delta^{(j)} \right) \\ &+ \gamma_5 e^z \left(\sum_{j=0}^{5k+m+1} \binom{5k+m+1}{j} F_\gamma^{(j)} \right), \end{aligned} \tag{4.44}$$

where Q_m is a differential polynomial in ϕ_5 and its derivatives with constant coefficients, such that each term of Q_m has at least one $\phi_5^{(l)}$, $l \geq k+1$, as a factor. Now choose $a_m \in \mathbb{R}$, where $m = 0, \dots, 6k+3$, such that the following system of linear equations is satisfied:

$$\left\{ \begin{array}{ll} \binom{5k+1}{0} a_0 + \binom{5k+2}{0} a_1 + \cdots + \binom{11k+4}{0} a_{6k+3} & = 0 \\ \binom{5k+1}{1} a_0 + \binom{5k+2}{1} a_1 + \cdots + \binom{11k+4}{1} a_{6k+3} & = 0 \\ \vdots & \vdots \\ \binom{5k+1}{3k} a_0 + \binom{5k+2}{3k} a_1 + \cdots + \binom{11k+4}{3k} a_{6k+3} & = 0 \\ 2^{5k+1} \binom{5k+1}{0} a_0 + 2^{5k+2} \binom{5k+2}{0} a_1 + \cdots + 2^{11k+4} \binom{11k+4}{0} a_{6k+3} & = 1 \\ 2^{5k} \binom{5k+1}{1} a_0 + 2^{5k+1} \binom{5k+2}{1} a_1 + \cdots + 2^{11k+3} \binom{11k+4}{1} a_{6k+3} & = 0 \\ 2^{5k-1} \binom{5k+1}{2} a_0 + 2^{5k} \binom{5k+2}{2} a_1 + \cdots + 2^{11k+2} \binom{11k+4}{2} a_{6k+3} & = 0 \\ \vdots & \vdots \\ 2^{2k+1} \binom{5k+1}{3k} a_0 + 2^{2k+2} \binom{5k+2}{3k} a_1 + \cdots + 2^{8k+4} \binom{11k+4}{3k} a_{6k+3} & = 0. \end{array} \right. \tag{4.45}$$

We may choose such a_m's, since all rows of (4.45) are linearly independent, and

(4.45) has more indeterminates than equations. Now, by (4.44), we have

$$
\begin{aligned}
\sum_{m=0}^{6k+3} a_m Q_m(z,\phi_5) = & \sum_{m=0}^{6k+3} \delta_5 e^{2z} \left(\sum_{j=0}^{5k+m+1} 2^{5k+m+1-j} \binom{5k+m+1}{j} a_m F_\delta^{(j)} \right) \\
& + \sum_{m=0}^{6k+3} \gamma_5 e^{z} \left(\sum_{j=0}^{5k+m+1} \binom{5k+m+1}{j} a_m F_\gamma^{(j)} \right) \\
& = \delta_5 e^{2z} \sum_{j=0}^{3k} \sum_{m=0}^{6k+3} 2^{5k+m+1-j} \binom{5k+m+1}{j} a_m F_\delta^{(j)} \\
& + \gamma_5 e^{z} \sum_{j=0}^{3k} \sum_{m=0}^{6k+3} \binom{5k+m+1}{j} a_m F_\gamma^{(j)} \\
& + \delta_5 e^{2z} \sum_{m=0}^{6k+3} \left(\sum_{j=3k+1}^{5k+m+1} 2^{5k+m+1-j} \binom{5k+m+1}{j} a_m F_\delta^{(j)} \right) \\
& + \gamma_5 e^{z} \sum_{m=0}^{6k+3} \left(\sum_{j=3k+1}^{5k+m+1} \binom{5k+m+1}{j} a_m F_\gamma^{(j)} \right),
\end{aligned}
$$

and so, by (4.45),

$$
\begin{aligned}
\delta_5 e^{2z} F_\delta = & - \delta_5 e^{2z} \sum_{m=0}^{6k+3} \left(\sum_{j=3k+1}^{5k+m+1} 2^{5k+m+1-j} \binom{5k+m+1}{j} a_m F_\delta^{(j)} \right) \\
& - \gamma_5 e^{z} \sum_{m=0}^{6k+3} \left(\sum_{j=3k+1}^{5k+m+1} \binom{5k+m+1}{j} a_m F_\gamma^{(j)} \right) + \sum_{m=0}^{6k+3} a_m Q_m(z,\phi_5).
\end{aligned}
$$

Hence,

$$
\begin{aligned}
m\left(r, \frac{1}{\phi_5^{(k+1)}}\right) & \le m\left(r, \frac{\delta_5 e^{2z} F_\delta}{\phi_5^{(k+1)}}\right) + m\left(r, \frac{1}{F_\delta}\right) + O(r) \\
& \le m\left(r, \frac{\delta_5 e^{2z} F_\delta}{\phi_5^{(k+1)}}\right) + 2m\left(r, \frac{1}{\phi_5}\right) + m\left(r, \frac{1}{\phi_5+1}\right) + O(r) \\
& = O(r).
\end{aligned}
$$

Thus

$$
m\left(r, \frac{1}{\phi_5^{(n)}}\right) = O(r) \tag{4.46}
$$

for all $n \in \mathbb{N} \cup \{0\}$ by the induction principle.

Suppose finally that $P \not\equiv 0$, and that $\deg(P) = p$. Let $j = 3, 5$. Then we have for all $n \in \mathbb{N} \cup \{0\}$,

$$m\left(r, \frac{1}{\phi_j^{(n)} - P}\right) \leq m\left(r, \frac{(\phi_j^{(n)} - P)^{(p+1)}}{\phi_j^{(n)} - P}\right) + m\left(r, \frac{1}{\phi_j^{(n+p+1)}}\right) = O(r),$$

by (4.28), (4.38) and (4.46). □

The following theorem gives similar information about the Valiron deficiencies of ϕ_3 and ϕ_5 as Theorem 4.2.3 does about the Valiron deficiencies of ϕ_1, ϕ_2 and ϕ_4.

Theorem 4.3.5. *Let $n \in \mathbb{N} \cup \{0\}$.*

1. *If $(\beta_3, \delta_3) \neq (0,0)$ and $(\alpha_3, \gamma_3) \neq (0,0)$, then $\Delta(P, \phi_3^{(n)}) = 0$ for every polynomial P.*
2. *If $\alpha_5\beta_5\delta_5 \neq 0$ and either $\gamma_5 \neq 0$, or $\alpha_5 \neq -\beta_5$, then $\Delta(P, \phi_5^{(n)}) = 0$ for every polynomial P.*
3. *If $(\alpha_3, \gamma_3) \neq (0,0)$, then $\Delta(\infty, \phi_3^{(n)}) = 0$.*
4. *If $\alpha_5 \neq 0$, then $\Delta(\infty, \phi_5^{(n)}) = 0$.*

We now use Lemma 4.3.4 to investigate the value distribution of $\phi_j^{(n)}$, $j = 3, 5$.

Theorem 4.3.6. *Let $(\beta_3, \delta_3) \neq (0,0)$ and $(\alpha_3, \gamma_3) \neq (0,0)$. Then*

$$\Phi_{(n)}(\phi_3, P) = \frac{n+2}{2}$$

for all $n \in \mathbb{N}$ if $\gamma_3 = 0$, and

$$\Phi_{(n)}(\phi_j, P) = n + 1 \tag{4.47}$$

for all $n \in \mathbb{N}$ if $\gamma_3 \neq 0$ and $j = 3$. Furthermore, if $\alpha_5\beta_5\delta_5 \neq 0$ and either $\gamma_5 \neq 0$, or $\alpha_5 \neq -\beta_5$, then (4.47) holds when $j = 5$.

Proof. Observe that when $\gamma_3 \neq 0$, all poles of ϕ_3 are simple. When $\gamma_3 = 0$, we have by assumption that $\alpha_3 \neq 0$, and so all poles are double in this case. The assertion now follows by completely similar reasoning as in the proof of Theorem 4.2.5 by using Lemma 4.3.4 instead of Lemma 4.2.2, and by remembering that we assumed ϕ_3 to be admissible. The reasoning of the proof is similar in the case $j = 5$. □

Also the proofs of the following three theorems are very similar to what we can find in Section 4.2. In general, we just have to apply Lemma 4.3.4, instead of Lemma 4.2.2, with slight modifications to the reasoning in each case. For this reason we omit all details of these proofs.

Theorem 4.3.7. *Let $n \in \mathbb{N}$, and let the assumptions of Theorem 4.3.6 hold. Then*

$$T(r, \phi_3^{(n)}) = \frac{n+2}{2} T(r, \phi_3) + O(r),$$

when $\gamma_3 = 0$, and

$$T(r, \phi_j^{(n)}) = (n+1) T(r, \phi_j) + O(r),$$

when $\gamma_3 \neq 0$ and $j = 3$, or when $j = 5$.

As we can see from Theorems 4.3.5 and 4.3.6, the value distribution of the derivatives of the meromorphic functions satisfying the modified third or the fifth Painlevé differential equation is also extremely regular, as was the case for the first, the second and the fourth Painlevé equations. The following theorem, which is analogous to Theorem 4.2.7, further strengthens this assertion.

Theorem 4.3.8. *Let $(\alpha_3, \gamma_3) \neq (0,0)$, $(\beta_3, \delta_3) \neq (0,0)$, $\alpha_5 \beta_5 \delta_5 \neq 0$, and either $\gamma_5 \neq 0$, or $\alpha_5 \neq -\beta_5$. Then for all $n \in \mathbb{N} \cup \{0\}$,*

$$\Phi_e(\phi_3^{(n)}) = \frac{n+3}{n+2}$$

if $\gamma_3 = 0$, and

$$\Phi_e(\phi_j^{(n)}) = \frac{n+2}{n+1}$$

for $j = 3$, if $\gamma_3 \neq 0$, and for $j = 5$.

We conclude this section by a version of the second main theorem for $\phi_j^{(n)}$, where $j = 3, 5$ and $n \in \mathbb{N} \cup \{0\}$.

Theorem 4.3.9. *Let $P_1, \ldots, P_q$ be distinct complex polynomials, and let $(\alpha_3, \gamma_3) \neq (0,0)$, $(\beta_3, \delta_3) \neq (0,0)$, $\alpha_5 \beta_5 \delta_5 \neq 0$, and either $\gamma_5 \neq 0$, or $\alpha_5 \neq -\beta_5$. Then for all $n \in \mathbb{N} \cup \{0\}$,*

$$\sum_{k=1}^{q} m\left(r, \frac{1}{\phi_3^{(n)} - P_k}\right) + N_B(r, \phi_3^{(n)}) = (n+2) T(r, \phi_3) + O(r),$$

if $\gamma_3 = 0$, and

$$\sum_{k=1}^{q} m\left(r, \frac{1}{\phi_j^{(n)} - P_k}\right) + N_B(r, \phi_j^{(n)}) = (2n+2) T(r, \phi_j) + O(r),$$

if $\gamma_3 \neq 0$ and $j = 3$, or if $j = 5$.

5. Difference-differential equations

In this section, we consider certain kinds of difference-differential equations, namely, equations which have finite order solutions with Borel exceptional zeros and poles. We consider the value distribution of such solutions, including an analogue for the lemma of the logarithmic derivative, and a Clunie type lemma. Finally, we add a few remarks about solutions permitting two arbitrary Borel exceptional slowly moving targets, instead of constant targets and infinity.

Let $\rho > 0$ be a given constant, and let

$$\mathcal{K}_\rho := \{h \text{ meromorphic} : \rho(h) < \rho\}.$$

The set $\mathcal{K}_\rho$ equipped with the operations of pointwise multiplication, addition, complex derivation and the independent variable transformation $z + c$, where $c \in \mathbb{C}$, forms a *difference-differential field.* This field is algebraically closed, in other words, any meromorphic function satisfying an algebraic equation over the field $\mathcal{K}_\rho$ belongs to $\mathcal{K}_\rho$. This follows by slightly modifying the proof of [14, Theorem 1].

We say that a meromorphic function f is *difference-differentially algebraic* with respect to a difference-differential field $\mathcal{K}_\rho$ if it satisfies an algebraic difference-differential equation

$$P(z, w) = 0, \tag{5.1}$$

where P is a *difference-differential polynomial* over $\mathcal{K}_\rho$, in other words, a polynomial in $f(z+c_1), \dots, f(z+c_k)$, $c_j \in \mathbb{C}$, and their derivatives with meromorphic coefficients $a_\lambda \in \mathcal{K}_\rho$.

5.1. Properties of some meromorphic functions of finite order

First, we need to prove some important lemmas concerning finite order meromorphic functions with Borel exceptional zeros and poles. We start with a new notation.

Definition 5.1.1. Let f be a meromorphic function of finite order ρ such that

$$\max\left(\lambda(f), \lambda\left(\frac{1}{f}\right)\right) < \rho. \tag{5.2}$$

The set of all functions satisfying the above condition is denoted by $\mathcal{F}_\rho$.

The following two lemmas makes dealing with the functions in $\mathcal{F}_\rho$ easier. The next lemma is in fact a corollary of the Hadamard theorem.

Lemma 5.1.2. *Let $f \in \mathcal{F}_\rho$. Then*

$$f(z) = z^m \frac{Q_g(z)}{Q_h(z)} e^{P(z)},$$

where $m \in \mathbb{Z}$, $P(z)$ is a polynomial of degree ρ and $Q_g(z)$ and $Q_h(z)$ are the canonical products formed with nonzero zeros and poles of f, respectively.

Proof. Take $\varepsilon > 0$. It is well known that there exist two entire functions g and h of finite order $\leq \rho + \varepsilon$ such that

$$f(z) = \frac{g(z)}{h(z)}.$$

By Theorem 1.1.4 we have,

$$f(z) = z^{m_g - m_h} \cdot \frac{Q_g(z)}{Q_h(z)} \cdot e^{P_g(z) - P_h(z)},$$

where the notations of Theorem 1.1.4 are used and the subscript refers to the corresponding entire function g or h. Moreover, $Q_g(z)$ is an entire function with zeros exactly at the nonzero zeros of f and, correspondingly, $Q_h(z)$ is an entire function with zeros exactly at the nonzero poles of f. From the assumption (5.2) and Lemma 1.1.5 it follows that $\exp(P_g(z) - P_h(z))$ must be exactly of order ρ. By denoting $m := m_g - m_h$ and $P(z) := P_g(z) - P_h(z)$, the assertion follows. □

Lemma 5.1.3. *Let $f \in \mathcal{F}_\rho$, and $k \in \mathbb{N} \cup \{0\}$. Then*

$$f^{(k)}(z + c) = \Omega_k(z) e^{P(z+c)},$$

where $P(z)$ is a polynomial of degree ρ, and $\rho(\Omega_k) < \rho$.

Proof. By Lemma 5.1.2,

$$f(z) = z^m \cdot \frac{Q_g(z)}{Q_h(z)} \cdot e^{P(z)}, \tag{5.3}$$

where $m \in \mathbb{Z}$, $P(z)$ is a polynomial of degree ρ, $Q_g(z)$ is an entire function with zeros exactly at the nonzero zeros of f, and, correspondingly, $Q_h(z)$ is an entire function with zeros exactly at the nonzero poles of f. We denote

$$\Psi(z) := z^m \cdot \frac{Q_g(z)}{Q_h(z)}$$

for convenience of notation. The case $k = 0$ follows directly by (5.3), since

$$\rho(\Psi) \le \max\left(\rho(Q_g), \rho(Q_h)\right) = \max\left(\lambda(f), \lambda\left(\frac{1}{f}\right)\right) < \rho.$$

Suppose that $k \ge 1$. By substituting $z + c$ into equation (5.3), and by differentiating, we have

$$f^{(k)}(z+c) = \Omega_k(z)e^{P(z+c)},$$

where Ω_k is a polynomial in $\Psi(z+c)$, $P'(z+c)$ and their derivatives with integer coefficients. We now consider the order of Ω_k. By Lemma 1.1.6 it follows that $\lambda(f(z)) = \lambda(f(z+c))$, and that $\lambda(1/f(z)) = \lambda(1/f(z+c))$. Thus, by (5.2) and Lemma 1.1.5,

$$\begin{aligned}\rho(\Psi(z+c)) &\le \max\left(\rho\left(Q_g(z+c)\right), \rho\left(Q_h(z+c)\right)\right)\\ &= \max\left(\lambda\left(Q_g(z+c)\right), \lambda\left(Q_h(z+c)\right)\right) < \rho\end{aligned}$$

for all $c \in \mathbb{C}$. Thus,

$$\begin{aligned}\rho\left(\Omega_k(z)\right) &\le \max_{j=0,\dots,k}\left(\rho(\Psi^{(j)}(z+c))\right)\\ &= \rho(\Psi(z+c)) < \rho.\end{aligned}$$

□

Remarks. 1. If $f \in \mathcal{F}_\rho$, then f is of regular growth by Lemma 1.1.7.

2. The set $\mathcal{F}_\rho$ does not form even a group with the operations of the usual addition or multiplication, since it does not contain the identity element with respect to either operation.

3. By Theorem 1.1.4 and Lemma 1.1.6 it follows that if $f(z) \in \mathcal{F}_\rho$ then also $f(z+c) \in \mathcal{F}_\rho$, where $c \in \mathbb{C}$.

The following two lemmas investigate the properties of sums and products of some elements of $\mathcal{F}_\rho$. They will be used below to consider solutions of certain difference equations.

Lemma 5.1.4. *Let $f \in \mathcal{F}_\rho$, and let $F(z) := \prod_{j=1}^{n} f(z+c_j)$, where $c_1, \dots, c_n \in \mathbb{C}$. Then $F \in \mathcal{F}_\rho$.*

Proof. We proceed to show that $\rho(F) = \rho(f)$, and that condition (5.2) holds for F. First, by Lemma 5.1.2,

$$F(z) = \prod_{j=1}^{n}(z+c_j)^m \frac{Q_g(z+c_j)}{Q_h(z+c_j)} e^{P(z+c_j)},$$

where $m \in \mathbb{Z}$, P is a polynomial of degree ρ, and Q_g, Q_h are the canonical products formed with the nonzero zeros and poles of f, respectively.

Let $(b_n)_{n\in\mathbb{N}}$ and $(\beta_n)_{n\in\mathbb{N}}$ be the zero sequences of $f(z)$ and $F(z)$, respectively. By Lemma 1.1.6 it follows that the sum $\sum_{n=1}^{\infty} |b_n|^{-\nu}$ converges, if and only if any of the sums $\sum_{n=1}^{\infty} |b_n - c_j|^{-\nu}$, $j = 1, \ldots, n$, converges. Since the sum of finite number of absolutely convergent series also converges, $\sum_{n=1}^{\infty} |\beta_n|^{-\nu}$ converges if $\sum_{n=1}^{\infty} |b_n|^{-\nu}$ converges. Hence $\lambda(F) \leq \lambda(f)$, and by similar reasoning also $\lambda(1/F) \leq \lambda(1/f)$. The inequalities can be strict, since there may be some cancellation of poles and zeros. However, it follows that

$$\max\left(\lambda(F), \lambda\left(\frac{1}{F}\right)\right) < \rho(f).$$

By denoting

$$G_j(z) := (z + c_j)^m \frac{Q_g(z + c_j)}{Q_h(z + c_j)},$$

we can see that F is of the form

$$F(z) = e^{na_\rho z^\rho + P_{\rho-1}(z)} \prod_{j=1}^{n} G_j(z),$$

where a_ρ is the leading coefficient of P and $P_{\rho-1}(z)$ is a polynomial of degree $\leq \rho - 1$. By condition (5.2) and Lemma 1.1.5, $\rho(G_j) < \rho$ for all $j = 1, \ldots, n$. Hence also

$$\rho\left(\prod_{j=1}^{n} G_j(z)\right) < \rho,$$

and so $\rho(F) = \rho$. □

Analogously to Lemma 5.1.4 we have the following result:

Lemma 5.1.5. *Let $f \in \mathcal{F}_\rho$, and let $F(z) := \sum_{j=1}^{n} f(z + c_j)$, where $c_1, \ldots, c_n \in \mathbb{C}$. Then either $F \in \mathcal{F}_\rho$, or $F \equiv 0$.*

Proof. First, by Lemma 5.1.2,

$$F(z) = \sum_{j=1}^{n} (z + c_j)^m \frac{Q_g(z + c_j)}{Q_h(z + c_j)} e^{P(z+c_j)},$$

where $m \in \mathbb{Z}$, P is a polynomial of degree ρ, and Q_g, Q_h are the canonical products formed with the nonzero zeros and poles of f, respectively. Now

$$F(z) = e^{a_\rho z^\rho} \sum_{j=1}^{n} (z + c_j)^m \frac{Q_g(z + c_j)}{Q_h(z + c_j)} e^{P_{\rho-1,j}(z)} =: e^{a_\rho z^\rho} G(z), \qquad (5.4)$$

where a_ρ is the leading coefficient of P, and $P_{\rho-1,j}$ is a polynomial of degree $\leq \rho - 1$ for $j = 1, \ldots, n$. If $G \equiv 0$ then also $F \equiv 0$, and the proof is complete in this case. Hence we may assume that $G \not\equiv 0$. Then

$$\begin{aligned} \lambda(F) = \lambda(G) \leq \rho(G) &= \rho\left(\sum_{j=1}^{n}(z+c_j)^m \frac{Q_g(z+c_j)}{Q_h(z+c_j)} e^{P_{\rho-1,j}(z)}\right) \\ &\leq \max_{j=1,\ldots,n} \rho\left((z+c_j)^m \frac{Q_g(z+c_j)}{Q_h(z+c_j)} e^{P_{\rho-1,j}(z)}\right) < \rho. \end{aligned} \tag{5.5}$$

The last inequality follows by (5.2) and Lemma 1.1.5. Similarly,

$$\lambda\left(\frac{1}{F}\right) = \lambda\left(\frac{1}{G}\right) < \rho.$$

We already saw in (5.5) that $\rho(G) < \rho$. Hence, by (5.4), $\rho(F) = \rho$, and so $F \in \mathcal{F}_\rho$ in this case. This completes the proof. □

The following example shows that both cases in Lemma 5.1.5 can occur.

Example 5.1.6. For $f(z) = e^z$, we have

$$f(z) + f(z + \pi i) \equiv 0,$$

and

$$f(z) + f(z + \pi) = e^z(1 + e^\pi).$$

Here

$$\rho\left(e^z(1 + e^\pi)\right) = \rho(f) = 1,$$

and $\lambda(e^z(1 + e^\pi)) = \lambda(1/e^z(1 + e^\pi)) = 0$.

Define the *type* of an entire function f of order ρ, such that $0 < \rho < \infty$, by

$$\tau(f) := \limsup_{r\to\infty} \frac{\log M(r, f)}{r^\rho},$$

where $M(r, f)$ denotes the maximum modulus of f on a disc $\{z : |z| \leq r\}$. For the proof of our next result, we need the following lemma, which is a special case of [20, Satz 11.3].

Lemma 5.1.7. *Let f_1 and f_2 be entire functions of order ρ such that $0 < \rho < \infty$. If $\tau(f_1) > \tau(f_2)$, then $\rho(f_1 f_2) = \rho(f_1)$, $\rho(f_1 + f_2) = \rho(f_1)$ and $\tau(f_1 + f_2) = \tau(f_1)$.*

We conclude this section with the following lemma, which, in a way, combines the assertions of Lemmas 5.1.4 and 5.1.5, although they are not used in the proof. This lemma has applications in the theory difference-differential equations, as we will see in Section 5.2.

Lemma 5.1.8. *Let $f \in \mathcal{F}_\rho$, and let*

$$F(z) := \sum_{\substack{\lambda=(j_1,\ldots,j_{k(n+1)})\in I \\ |\lambda|\geq 1}} a_\lambda(z) \prod_{m=1}^{k}\prod_{l=0}^{n} f^{(l)}(z+c_m)^{j_{l+1+(m-1)(n+1)}},$$

where $c_1, \ldots, c_k \in \mathbb{C}$, $n \in \mathbb{N} \cup \{0\}$, and $a_\lambda \in \mathcal{K}_\rho$ for all $\lambda \in I$. Then either $\rho(F) = \rho$, or $F \equiv 0$.

Proof. Suppose that $F \not\equiv 0$. By Lemma 5.1.3 we have for all $l \in \mathbb{N} \cup \{0\}$,

$$f^{(l)}(z+c) = \Omega_l(z)e^{P(z+c)},$$

where $P(z)$ is a polynomial of degree ρ and $\rho(\Omega_l) < \rho$. Hence,

$$\begin{aligned} F(z) &= \sum_{\substack{\lambda\in I \\ |\lambda|\geq 1}} b_\lambda(z) e^{(j_1+j_2+\cdots+j_{n+1})P(z+c_1)} \cdots e^{(j_{(n+1)(k-1)+1}+\cdots+j_{k(n+1)})P(z+c_k)} \\ &= \sum_{\substack{\lambda\in I \\ |\lambda|\geq 1}} b_\lambda(z) e^{J_1P(z+c_1)+J_2P(z+c_2)+\cdots+J_kP(z+c_k)}, \end{aligned}$$

where $b_\lambda \in \mathcal{K}_\rho$ for all $\lambda \in I$, and

$$J_l = j_{(n+1)(l-1)+1} + j_{(n+1)(l-1)+2} + \cdots + j_{(n+1)(l-1)+n+1}$$

for $l = 1, \ldots, k$. Let $A(z)$ be the canonical product formed with the poles of all $b_\lambda(z)$, $\lambda \in I$, counting multiplicities. Then clearly $A \in \mathcal{K}_\rho$, AF is entire, and

$$A(z)F(z) = \sum_{j=1}^{N} e^{ja_\rho z^\rho} \left(\sum_{s=1}^{\sigma(j)} C_{j,s}(z)e^{P_{j,s}(z)} \right),$$

where $P_{j,s}(z)$ are polynomials of degree $\leq \rho-1$, $C_{j,s}(z) \in \mathcal{K}_\rho$ are entire functions, and N is the greatest integer such that $\sum_{s=1}^{\sigma(N)} C_{N,s}(z)e^{P_{N,s}(z)} \not\equiv 0$. Clearly such N exists, since otherwise $F(z) \equiv 0$, which is a contradiction to our assumption. Now,

$$A(z)F(z) = e^{Na_\rho z^\rho}G(z), \tag{5.6}$$

where

$$G(z) := \sum_{j=1}^{N} e^{(j-N)a_\rho z^\rho} \left(\sum_{s=1}^{\sigma(j)} C_{j,s}(z)e^{P_{j,s}(z)} \right). \tag{5.7}$$

Now we have the following three possibilities:

1. $G(z) \equiv 0$. Then by (5.6) also $A(z)F(z) \equiv 0$, and hence $F(z) \equiv 0$, since clearly $A(z) \not\equiv 0$. This contradicts our assumption.

Corollary 5.2.3. *Let $f \in \mathcal{F}_\rho$. Then*

$$\begin{array}{ll} (i) & m\left(r, \frac{f^{(k)}(z+c)}{f(z)}\right) = o(T(r,f)), \\ (ii) & N\left(r, \frac{f^{(k)}(z+c)}{f(z)}\right) = o(T(r,f)), \\ (iii) & T\left(r, \frac{f(z)}{f^{(k)}(z+c)}\right) = o(T(r,f)), \\ (iv) & m\left(r, \frac{f(z)}{f^{(k)}(z+c)}\right) = o(T(r,f)), \\ (v) & N\left(r, \frac{f(z)}{f^{(k)}(z+c)}\right) = o(T(r,f)) \end{array}$$

for all $c \in \mathbb{C}$ and $k \in \mathbb{N} \cup \{0\}$.

Proof. Since by Lemma 5.2.2,

$$T\left(r, \frac{f^{(k)}(z+c)}{f(z)}\right) = N\left(r, \frac{f^{(k)}(z+c)}{f(z)}\right) + m\left(r, \frac{f^{(k)}(z+c)}{f(z)}\right) = o(T(r,f)),$$

and both $m\left(r, f^{(k)}(z+c)/f(z)\right)$ and $N\left(r, f^{(k)}(z+c)/f(z)\right)$ are non-negative for r large enough, we have (i) and (ii). Since (iii) follows by the first main theorem and Lemma 5.2.2, we can repeat the above reasoning to prove (iv) and (v). □

The following three examples show that the condition $f \in \mathcal{F}_\rho$ is necessary. The first example demonstrates that Lemma 5.2.2 is not true in general for functions of infinite order.

Example 5.2.4. Consider the function

$$f(z) := e^{e^z}.$$

Then

$$\begin{aligned} T\left(r, \frac{f(z+1)}{f(z)}\right) &= m\left(r, \frac{f(z+1)}{f(z)}\right) = (e-1)m(r,f) \\ &= (e-1)T(r,f) \neq o(T(r,f)). \end{aligned}$$

The next example shows that we can not omit the condition (5.2).

Example 5.2.5. Consider first the function

$$f(z) := \cos(z).$$

Then clearly $\lambda(f) = \rho(f) = 1$. Moreover, by a simple calculation,

$$T(r,f) = \frac{2r}{\pi} + O(1),$$

and similarly,

$$T(r, \tan(z)) = \frac{4r}{\pi} + O(1).$$

But now

$$T\left(r, \frac{f\left(z-\frac{\pi}{2}\right)}{f(z)}\right) = T\left(r, \tan(z)\right) = \frac{4r}{\pi} + O(1) \neq o(T(r,f)).$$

Similarly by considering the function

$$f(z) := \frac{1}{\cos(z)},$$

we obtain a counterexample for the case $\lambda(1/f) = \rho(f) = 1$.

Finally, we demonstrate that 0 and ∞ are essential in (5.2).

Example 5.2.6. The function $f(z) := \tan(z)$ has the Borel exceptional values $\pm i$, but

$$T\left(r, \frac{f'}{f}\right) = T\left(r, \frac{1+\tan^2(z)}{\tan(z)}\right) = 2T(r, \tan(z)) + O(1) \neq o(T(r,f)).$$

In what follows, we denote by $\deg(P)$ the total degree of difference-differential polynomial $P(z,f)$ as a polynomial in $f(z+c_1), \ldots, f(z+c_k)$ and their derivatives.

The following theorem is an interesting application of Lemma 5.2.1. This theorem will be used below to obtain a Clunie-type lemma for difference-differential equations. Note that the definition of a *homogeneous form* of a difference-differential polynomial is completely parallel to the definition of the homogeneous form of a differential polynomial, see Definition 2.3.1.

Theorem 5.2.7. *Let $f \in \mathcal{F}_\rho$, and let*

$$P(z,u) := \sum_{\lambda=(j_1,\ldots,j_{k(n+1)})\in I} a_\lambda(z) \prod_{m=1}^{k} \prod_{l=0}^{n} u^{(l)}(z+c_m)^{j_{l+1+(m-1)(n+1)}},$$

where $c_1, \ldots, c_k \in \mathbb{C}$ and $a_\lambda \in \mathcal{K}_\rho$ for all $\lambda \in I$. If $P(z,f) \equiv 0$, then $P(z,u)$ is a homogeneous form, and for each homogeneous component, $P_j(z,f) \equiv 0$.

Proof. By Lemma 5.1.8, and since we assumed $P(z,f) \equiv 0$, either $P(z,0) = P(z,0) - P(z,f) \equiv 0$, or $\rho(P(z,0)) = \rho(P(z,0) - P(z,f)) = \rho$. Since clearly $\rho(P(z,0)) < \rho$, we must have $P(z,0) \equiv 0$.

Next denote

$$P(z,f) := \sum_{j=1}^{\deg(P)} H_j(z,f),$$

where H_j is a homogeneous component of P of degree j for all $j = 1, \ldots, \deg(P)$. Assume that $H_l(z,f) \not\equiv 0$ for some $l = 1, \ldots, \deg(P)$. Assume further that l is the smallest such number. Then f is a solution of

$$H_l(z,u) = -\sum_{j=l+1}^{\deg(P)} H_j(z,u),$$

and so also a solution of

$$\frac{H_l(z,f)}{f^l} = -\sum_{j=l+1}^{\deg(P)} \frac{H_j(z,u)}{u^l}. \tag{5.11}$$

Now, $\rho(H_l/f^l) < \rho$ by Lemma 5.2.1, and $H_l/f^l \not\equiv 0$ by assumption. On the other hand, by Lemma 5.1.8, either

$$\rho\left(-\sum_{j=l+1}^{\deg(P)} \frac{H_j(z,f)}{f^l}\right) = \rho,$$

or

$$-\sum_{j=l+1}^{\deg(P)} \frac{H_j(z,f)}{f^l} \equiv 0.$$

Both cases lead to a contradiction. Hence $H_j(z,f) \equiv 0$ for all $j = 1, \ldots, \deg(P)$, and the proof is complete. □

Remark. Note that if $c_j = 0$ for all $j = 1, \ldots, k$, then Theorem 5.2.7 reduces into a special case of Lemma 2.3.2.

Theorem 5.2.7 gives a powerful tool for handling such difference-differential equations, which have a solution in $\mathcal{F}_\rho$. The following Clunie type lemma is an example of a direct application of Theorem 5.2.7.

Lemma 5.2.8. *Let $f \in \mathcal{F}_\rho$ be a transcendental meromorphic solution of*

$$f^n P(z,f) = Q(z,f), \tag{5.12}$$

where $P(z,f)$ and $Q(z,f)$ are polynomials in $f(z+c_1), \ldots, f(z+c_k)$, $c_j \in \mathbb{C}$, and their derivatives with meromorphic coefficients $\{a_\lambda \in \mathcal{K}_\rho : \lambda \in I\}$. Then

(i) *$P(z,f) \equiv 0$, if $\deg(Q) < n$, and*

(ii) $P(z, f) \equiv P(z, 0)$, *if* $\deg(Q) = n$.

Proof. Consider equation (5.12) in the form

$$f^n P(z, f) - Q(z, f) = 0. \tag{5.13}$$

(i) Suppose first that $\deg(Q) < n$. Then by Theorem 5.2.7 each homogeneous component of (5.13) must vanish identically. Hence $Q(z, f) \equiv 0$, and so also $P(z, f) \equiv 0$, since $f \not\equiv 0$.

(ii) Suppose now that $\deg(Q) = n$. Let Q_n be the homogeneous component of Q of degree n. Then by Theorem 5.2.7,

$$f^n P(z, 0) = Q_n(z, f), \tag{5.14}$$

since $f^n P(z, 0) - Q_n(z, f)$ is a homogeneous component of the equation (5.13). Moreover, by Theorem 5.2.7, all homogeneous components of (5.13), with degree strictly less than n, must vanish identically. Hence by (5.12) and (5.14),

$$P(z, f) = \frac{Q(z, f)}{f^n} = \frac{Q_n(z, f)}{f^n} = P(z, 0),$$

and the proof is complete. □

The following theorem is analogous to Theorem 2.3.4.

Theorem 5.2.9. *Let f be a meromorphic function of finite order ρ, and let $P(z, u)$ be a difference-differential polynomial with coefficients in the field $\mathcal{K}_\rho$, such that $P(z, f) \equiv 0$.*

1. *If there exist $g_j \in \mathcal{K}_\rho$, $j = 1, 2$, such that*

$$\lambda(f - g_j) < \rho,$$

then there exists a Möbius transformation Φ with coefficients in $\mathcal{K}_\rho$, and a homogeneous form Q over $\mathcal{K}_\rho$, such that $Q(z, \Phi(f)) \equiv 0$.

2. *If $\lambda(1/f) < \rho$, and there exists $g \in \mathcal{K}_\rho$ such that*

$$\lambda(f - g) < \rho,$$

then there exists a Möbius transformation Φ with coefficients in $\mathcal{K}_\rho$, and a homogeneous form Q over $\mathcal{K}_\rho$, such that $Q(z, \Phi(f)) \equiv 0$.

Proof. If P is a homogeneous form, just take $Q \equiv P$ and $\Phi(f) \equiv f$. Hence we may assume that P is nonhomogeneous. By assumption there exist two functions $g_1, g_2 \in \mathcal{K}_\rho$ such that

$$\lambda(f - g_j) < \rho$$

for $j = 1, 2$. Clearly the function

$$\Phi(f) := \frac{f - g_1}{f - g_2}$$

is difference-differentially algebraic over $\mathcal{K}_\rho$ satisfying, say, an algebraic difference-differential equation $Q(z, u) = 0$. Clearly also

$$\max(\lambda(\Phi(f)), \lambda(1/\Phi(f))) < \rho = \rho(\Phi(f)).$$

Now, Q is a homogeneous form by Theorem 5.2.7.

We still have to consider the case $\lambda(1/f) < \rho$. Assume that there exist $g \in \mathcal{K}_\rho$ such that

$$\lambda(f - g) < \rho,$$

and denote $\Phi(f) := f - g$. The rest of the proof follows by a completely similar reasoning as was done above. □

Remark. If again $c_1, \ldots, c_k = 0$, Theorem 5.2.9 reduces into a special case of Theorem 2.3.4.

5.3. An application

In this section, we use the earlier obtained results to study the properties of some difference equations having solutions in $\mathcal{F}_\rho$.

The following theorem was proved in [13].

Theorem 5.3.1. *Suppose y is a transcendental meromorphic solution of*

$$\prod_{j=1}^{n} y(z + c_j) = \frac{a_0(z) + a_1(z)y + \cdots + a_p(z)y^p}{b_0(z) + b_1(z)y + \cdots + b_q(z)y^q}, \tag{5.15}$$

with coefficients $a_j(z), b_j(z) \in \mathcal{L}_y$ normalized such that $b_q = 1$. If

$$\max(\lambda(y), \lambda(1/y)) < \rho(y), \tag{5.16}$$

then (5.15) is either of the form

$$\prod_{j=1}^{n} y(z + c_j) = a_p(z)y(z)^p, \tag{5.17}$$

or

$$\prod_{j=1}^{n} y(z + c_j) = \frac{a_0(z)}{y(z)^q}. \tag{5.18}$$

We use Lemma 5.1.4 to improve this theorem in the finite order case.

Theorem 5.3.2. *Suppose $y \in \mathcal{F}_\rho$ is a transcendental meromorphic solution of* (5.15) *with coefficients $a_j(z), b_j(z) \in \mathcal{K}_\rho$ normalized such that $b_q = 1$. Then equation* (5.15) *reduces to the form*

$$\prod_{j=1}^{n} y(z+c_j) = a_n(z)y(z)^n. \tag{5.19}$$

Proof. By Theorem 5.3.1, (5.15) is either of the form (5.17) or (5.18). Consider first the case (5.18). By multiplying out the denominator, we have

$$y(z)^q \prod_{j=1}^{n} y(z+c_j) = a_0(z). \tag{5.20}$$

By Lemma 5.1.4 the left hand side of (5.20) is of order ρ. This contradicts the fact that $a_0 \in \mathcal{K}_\rho$.

Suppose now that (5.15) is of the form (5.17) and assume $p < n$. Then we obtain an equation of the form

$$\prod_{j=p+1}^{n} y(z+c_j) = a_p(z) \prod_{j=1}^{p} \frac{y(z)}{y(z+c_j)}. \tag{5.21}$$

The left hand side of (5.21) is of order ρ, again by Lemma 5.1.4. On the other hand, by Lemma 5.2.1, the right hand side of (5.21) belongs to the field $\mathcal{K}_\rho$, which is a contradiction. Since the case $p > n$ leads to a contradiction by a similar reasoning, we conclude that $p = n$. □

The following example demonstrates that the case (5.19) can occur.

Example 5.3.3. The equation

$$y(z+1)y(z-1) = e^2 y(z)^2$$

has a solution $f(z) = e^{z^2}$.

References

[1] Ablowitz, M. J., R. Halburd, and B. Herbst: On the extension of the Painlevé property to difference equations. - Nonlinearity 13, 2000, 889–905.

[2] Bank, S., and R. Kaufman: A note on Hölder's theorem concerning the Gamma function. - Math. Ann. 232, 1978, 115–120.

[3] BERGWEILER, W., K. ISHIZAKI, and N. YANAGIHARA: Meromorphic solutions of some functional equations. - Methods Appl. Anal. 5(3), 1998, 248–258.

[4] CARROLL, F. W.: Transcendental transcendence of solutions of Schröder's equation associated with finite Blaschke products. - Michigan Math. J. 32, 1985, 47–57.

[5] CHEVALLEY, C.: Introduction to the theory of algebraic functions of one variable. - Amer. Math. Soc., Baltimore, 1951.

[6] FRANK, G., and Y. WANG: On the meromorphic solutions of algebraic differential equations. - Analysis 18(1), 1998, 49–54.

[7] GOLUBEW, W.: Differentialgleichungen im Komplexen. - VEB Deutscher Verlag der Wissenschaften, Berlin, 1958.

[8] GUNDERSEN, G.: Finite order solutions of second order linear differential equations. - Trans. Amer. Math. Soc. 305(1), 1988, 415–429.

[9] GUNDERSEN, G.: On the real zeros of solutions of $f'' + A(z)f = 0$ where $A(z)$ is entire. - Ann. Acad. Sci. Fenn. 11, 1986, 275–294.

[10] GUNDERSEN, G.: The logarithmic derivative of a meromorphic function. - J. London Math. Soc. 37(2), 1988, 88–104.

[11] HAYMAN, W. K.: Meromorphic functions. - Clarendon Press, Oxford, 1964.

[12] HAYMAN, W. K.: The growth of solutions of algebraic differential equations, Atti. Acad. Naz. Lincei Cl. Sci. Fis. Mat. Natur. Rend. Lincei (9) Mat. Appl. 7(2), 1996, 67–73.

[13] HEITTOKANGAS, J., R. KORHONEN, I. LAINE, J. RIEPPO, and K. TOHGE: Complex difference equations of Malmquist type. - CMFT 1, 2001, 27–39.

[14] HELLERSTEIN, S., and L. A. RUBEL: Subfields that are algebraically closed in the field of meromorphic functions. - J. Anal. Math. 12, 1964, 105–111.

[15] HINKKANEN, A., and I. LAINE: Solutions of a modified fifth Painlevé equation are meromorphic. - Papers on analysis, Rep. Univ. Jyväskylä Dep. Math. Stat. 83, 2001, 133–146.

[16] HINKKANEN, A., and I. LAINE: Solutions of a modified third Painlevé equation are meromorphic. - J. Anal. Math. 85, 2001, 323–337.

[17] HINKKANEN, A., and I. LAINE: Solutions of the first and second Painlevé equations are meromorphic. - J. Anal. Math. 79, 1999, 345–377.

[18] ISHIZAKI, K.: Deficient and ramified small functions for admissible meromorphic solutions of some differential equations. - Complex Variables 13, 1990, 173–183.

[19] JACOBSON, N.: Lectures in abstract algebra. - D. Van Nostrand Company, Inc., New Jersey, 1966.

[20] JANK, G., and L. VOLKMANN, Einführung in die Theorie der ganzen und meromorphen Funktionen mit Anwendungen auf Differentialgleichungen. - Birkhäuser, Basel, 1985.

[21] LAINE, I.: Nevanlinna theory and complex differential equations. - Walter de Gruyter, Berlin, 1993.

[22] MOHON'KO, A. Z., and V. D. MOHON'KO: Estimates for the Nevanlinna characteristic of some classes of meromorphic functions and their applications to differential equations. - Sibirsk. Mat. Zh. 15, 1974, 1305–1322 [Russian]. Engl. Trans.: Siberian Math. J. 15, 1974, 921–934.

[23] MURATA, Y.: Rational solutions of the second and the fourth Painlevé equations. - Funkcial. Ekvac. 28, 1985, 1–32.

[24] PAINLEVÉ, P.: Leçons sur la théorie analytique des équations différentielles, profesées à Stockholm, 1895. - Hermann, Paris, 1897.

[25] PICARD, E.: Mémoire sur la théorie des fonctions algébriques de deux variables. - J. de Math. 5(4), 1889, 135–319.

[26] SHIMOMURA, S.: Growth of modified Painlevé transcendents of the third and the fifth kind. - Preprint.

[27] SHIMOMURA, S.: Growth of the first, second and the fourth Painlevé transcendents. - Preprint.

[28] SHIMOMURA, S.: On deficiences and ramification indices for Painlevé transcendents of the second kind. - Preprint.

[29] SHIMOMURA, S.: On deficiences of small functions for Painlevé transcendents of the fourth kind. - Ann. Acad. Sci. Fenn. 27, 2002, 109–120.

[30] SHIMOMURA, S.: Value distribution of Painlevé transcendents of the fifth kind. - Result. Math. 38, 2000, 348–361.

[31] SHIMOMURA, S.: Value distribution of Painlevé transcendents of the first and the second kind. - J. Anal. Math. 82, 2000, 333–346.

[32] SHIMOMURA, S.: Value distribution of Painlevé transcendents of the third kind, Complex Variables 40, 1999, 51–62.

[33] STEINMETZ, N.: Eigenschaften eindeutiger Lösungen gewöhnlicher Differentialgleichungen im Komplexen. - Karlsruhe, 1978.

[34] STEINMETZ, N.: On Painlevé's equations I, II and IV, J. Anal. Math. 82, 2000, 363–377.

[35] STEINMETZ, N.: Value distribution of the Painlevé transcendents. - Israel J. Math. 128, 2002, 29–52.

[36] YANG LO: Precise estimate of total deficiency of meromorphic derivatives. - J. d'Analyse Math. 55, 1990, 287–296.

[37] YANG LO: Value distribution theory. - Springer-Verlag, Berlin, 1993.

Appendix A: The coefficients in Theorem 3.2.2

Since the explicit forms of the coefficient functions in Theorem 3.2.2 turned out to be extremely complicated in most cases, we present them here. They were calculated by Maple© mathematical software. The coefficients A_k, $k = 1, \ldots, 6$ are:

$$\begin{aligned}
A_1 &= c(h_1 - h_2), && \text{(A.1)}\\
A_2 &= -4ah_2h_1^3 + 2h_2'h_1' - 2ah_1^3b + 2ah_1bc_1c_2 && \text{(A.2)}\\
&\quad +ab^2c_1h_1 - 2ab^2c_1c_2 + ab^2h_1c_2 + 2ah_1^4 + 2ah_2bc_1c_2 + ab^2h_2c_2 + ab^2c_1h_2\\
&\quad -2ah_2h_1c_1c_2 - 4ah_2bh_1c_2 - 4ah_2bc_1h_1 - 2ab^2h_2h_1\\
&\quad +3ah_2h_1^2c_1 + 6ah_2bh_1^2 + 3ah_2h_1^2c_2 - ah_1^3c_2 - ah_1^3c_1,\\
A_3 &= -ah_1^3c_3 - ah_1^3c_4 - ac_1h_1^3 - ah_1^3c_2 - 4ah_2h_1^3 + 2ah_1^4 + ac_1h_1c_3c_4 && \text{(A.3)}\\
&\quad +ah_1c_2c_3c_4 + ac_1c_2h_1c_4 + ac_1c_2c_3h_1 - 2ac_1c_2c_3c_4 + 3ah_1^2c_3h_2\\
&\quad +3ah_1^2c_2h_2 + 3ac_1h_1^2h_2 + 3ah_2h_1^2c_4 - 2ac_1h_1c_3h_2 - 2ah_1c_2c_3h_2 - 2ac_1c_2h_1h_2\\
&\quad +ac_1c_2c_3h_2 + 2h_2'h_1' + ac_1c_2h_2c_4 - 2ah_2h_1c_3c_4 - 2ah_2c_2h_1c_4\\
&\quad +ah_2c_2c_3c_4 - 2ac_1h_2h_1c_4 + ac_1h_2c_3c_4,\\
A_4 &= -3ah_1^6 + 4ac_1^2h_2h_1^3 - 10ah_1^4c_2h_2 - 10ah_1^4h_2c_3 + 4ah_2h_1^3c_3^2 && \text{(A.4)}\\
&\quad +ac_1^2c_2^2h_1^2 - 4ah_1^4c_1c_3 + ac_1^2h_1^2c_3^2 + ah_1^2c_2^2c_3^2 - ah_1^4c_3^2 + 4ah_1^5c_3\\
&\quad -ah_1^4c_2^2 + 4ah_1^5c_2 + 4ah_1^5c_1 + 6ah_2h_1^5 - ac_1^2h_1^4 - 10ah_2c_1h_1^4 + 4ah_2h_1^3c_2^2\\
&\quad +8ac_1^2h_2c_2h_1c_3 - 2ac_1^2h_2c_2c_3^2 - 6ah_2h_1^2c_2c_3^2 - 6ah_2c_1c_2^2h_1^2\\
&\quad +16ah_1^3c_2h_2c_3 - 6ah_2h_1^2c_2^2c_3 + 2ac_1^2h_2h_1c_3^2 - 3h_2'(h_1')^2 + 2ah_2h_1c_2^2c_3^2\\
&\quad -6ac_1^2h_2c_2h_1^2 + 16ah_2c_1h_1^3c_2 - 6ah_2c_1h_1^2c_3^2 + 8ah_2c_1c_2^2h_1c_3\\
&\quad +16ah_2c_1h_1^3c_3 - 2ah_2c_1c_2^2c_3^2 - 24ah_2c_1h_1^2c_2c_3 + 8ah_2c_1h_1c_2c_3^2\\
&\quad -6ac_1^2h_1^2h_2c_3 + 2ac_1^2c_2^2h_2h_1 - 2ac_1^2c_2^2h_2c_3 - 4ah_1^4c_2c_3 - 4ah_1^4c_1c_2\\
&\quad +4ac_1^2h_1^2c_2c_3 + 3ac_1^2c_2^2c_3^2 - 4ac_1^2h_1c_2c_3^2 - 4ah_1c_1c_2^2c_3^2\\
&\quad -4ac_1^2c_2^2h_1c_3 + 4ah_1^2c_1c_2c_3^2 + 4ah_1^2c_1c_2^2c_3,\\
A_5 &= 9ah_1^7c_2 + 6ah_1^7c_1 - 4h_2'(h_1')^3 - 6ah_1^6c_2^2 + 8ah_2h_1^7 + ah_1^5c_3^3 && \text{(A.5)}\\
&\quad +9ah_1^7c_3 - 2ac_1^2h_1^6 - 6ah_1^6c_3^2 + ah_1^5c_2^3 - 5ah_2h_1^4c_3^3 + 9ah_1^5c_2^2c_3\\
&\quad -3ah_1^3c_2^2c_3^3 + 2ah_1^2c_2^3c_3^3 - ac_1^2h_1^3c_3^3 + 4ac_1^2c_2^3c_3^3 - 3ah_1^3c_2^3c_3^2
\end{aligned}$$

$$
\begin{aligned}
& -12ah_1^6c_1c_3 + 6ah_1^5c_1c_3^2 + 3ac_1^2h_1^5c_3 - 12ah_1^6c_1c_2 + 3ac_1^2h_1^5c_2 \\
& +6ah_1^5c_1c_2^2 - 18ah_1^6c_2c_3 + 9ah_1^5c_2c_3^2 - ac_1^2c_2^3h_1^3 - 4ah_1^8 + 18ah_1^5c_1c_2c_3 \\
& -18ah_1^3c_1c_2^2c_3^2 - 6ah_1^3c_1c_2^3c_3 + 12ah_1^2c_1c_2^3c_3^2 - 9ac_1^2h_1^3c_2c_3^2 \\
& -9ac_1^2h_1^3c_2^2c_3 + 18ac_1^2h_1^2c_2^2c_3^2 + 6ac_1^2c_2^3h_1^2c_3 - 9ac_1^2c_2^3h_1c_3^2 \\
& -6ah_1^3c_1c_2c_3^3 + 12ah_1^2c_1c_2^2c_3^3 - 6ah_1c_1c_2^3c_3^3 + 6ac_1^2h_1^2c_2c_3^3 \\
& -9ac_1^2h_1c_2^2c_3^3 - 27ac_1^2h_1^2c_2h_2c_3^2 - 45ah_1^4c_2h_2c_3^2 + 36ah_1^3c_2^2h_2c_3^2 \\
& -9ah_1^2c_2^3h_2c_3^2 - 30ah_1^4c_1h_2c_3^2 + 12ac_1^2h_1^3h_2c_3^2 - 3ac_1^2c_2^3h_2c_3^2 \\
& +12ah_1c_1c_2^3h_2c_3^2 - 27ac_1^2h_2c_2^2h_1^2c_3 + 18ac_1^2h_2c_2^2h_1c_3^2 \\
& -3ac_1^2h_2c_2^2c_3^3 - 18ah_2c_1h_1^2c_2c_3^3 + 12ah_2c_1h_1c_2^2c_3^3 \\
& +36ah_1^5c_1c_2h_2 - 90ah_2c_1h_1^4c_2c_3 + 72ah_2c_1h_1^3c_2c_3^2 + 72ah_2c_1h_1^3c_2^2c_3 \\
& -54ah_2c_1h_1^2c_2^2c_3^2 - 3ac_1^2h_2h_1^2c_3^3 + 6ac_1^2h_2h_1c_2c_3^3 + 12ah_2h_1^3c_2c_3^3 \\
& -9ah_2h_1^2c_2^2c_3^3 + 2ah_2h_1c_2^3c_3^3 + 8ah_2c_1h_1^3c_3^3 - 2ah_2c_1c_2^3c_3^3 - 21ah_2h_1^6c_3 \\
& +18ah_2h_1^5c_3^2 - 15ac_1^2h_1^4c_2h_2 + 12ac_1^2h_1^3c_2^2h_2 - 3ac_1^2c_2^3h_2h_1^2 \\
& -21ah_1^6c_2h_2 + 18ah_1^5c_2^2h_2 - 5ah_1^4c_2^3h_2 - 14ah_1^6c_1h_2 + 6ac_1^2h_1^5h_2 \\
& -30ah_1^4c_1c_2^2h_2 + 8ah_1^3c_1c_2^3h_2 + 54ah_1^5c_2h_2c_3 - 45ah_1^4c_2^2h_2c_3 \\
& +12ah_1^3c_2^3h_2c_3 + 36ah_1^5c_1h_2c_3 - 15ac_1^2h_1^4h_2c_3 + 6ac_1^2c_2^3h_2h_1c_3 \\
& +36ac_1^2h_1^3c_2h_2c_3 - 18ah_1^2c_1c_2^3h_2c_3,
\end{aligned}
$$

$$
\begin{aligned}
A_6 = {} & -25ah_1^11c_3 + 6(h_1')^5h_2' + 2ah_1^8c_2^4 - 10ah_1^4c_1^3c_2^4c_3 - 60ah_1^7c_1c_2^3c_3 \qquad (A.6) \\
& +ah_1^5c_1^3c_2^4 - 30ah_1^9c_3^3 - 90ah_1^7c_1^2c_2^2c_3 - 12ah_1^9c_2^3 - 120ah_1^7c_1c_2c_3^3 \\
& -180ah_1^7c_1c_2^2c_3^2 - 180ah_1^9c_1c_2c_3 + 6ah_1^12 - 120ah_1^7c_1^2c_2c_3^2 \\
& -12ah_1^2c_1c_2^4c_3^5 + 25ah_1c_1^3c_2^4c_3^4 + 15ah_1c_1^2c_2^4c_3^5 + 120ah_1^8c_1^2c_2c_3 \\
& -6ac_1^3c_2^4c_3^5 + 240ah_1^8c_1c_2c_3^2 + 180ah_1^8c_1c_2^2c_3 + 20ah_1c_1^3c_2^3c_3^5 \\
& +180ah_1^3c_1^2c_2^3c_3^4 + 120ah_1^3c_1^3c_2^3c_3^3 + 54ah_1^3c_1^2c_2^2c_3^5 \\
& +45ah_1^3c_1c_2^4c_3^4 + 30ah_1^3c_1^3c_2^4c_3^2 + 12ah_1^3c_1^3c_2c_3^5 - 24ah_1^4c_1^2c_2c_3^5 \\
& +90ah_1^3c_1^3c_2^2c_3^4 + 90ah_1^3c_1^2c_2^4c_3^3 + 36ah_1^3c_1c_2^3c_3^5 - 10ah_1^4c_2^4c_3^4 \\
& -80ah_1^2c_1^3c_2^3c_3^4 - 40ah_1^4c_1^3c_2c_3^4 - 240ah_1^4c_1^2c_2^3c_3^3 - 120ah_1^4c_1^3c_2^2c_3^3 \\
& -60ah_1^4c_1c_2^4c_3^3 + 60ah_1^5c_1^2c_2c_3^4 - 8ah_1^7h_2c_2^4 - 40ah_1^2c_1^3c_2^4c_3^3 \\
& +180ah_1^5c_1^2c_2^2c_3^3 + 40ah_1^5c_1^3c_2c_3^3 + 20ah_1^5c_1^3c_2^3c_3 + 15ah_1^5c_1^2c_2^4c_3 \\
& +30ah_1^5c_1c_2^4c_3^2 + 120ah_1^5c_1c_2^3c_3^3 + 90ah_1^5c_1c_2^2c_3^4 + 12ah_1^5c_1c_2c_3^5 \\
& +120ah_1^5c_1^2c_2^3c_3^2 + 60ah_1^5c_1^3c_2^2c_3^2 - 180ah_1^4c_1^2c_2^2c_3^4 - 80ah_1^4c_1^3c_2^3c_3^2 \\
& -36ah_1^4c_1c_2^2c_3^5 - 120ah_1^4c_1c_2^3c_3^4 - 24ah_1^2c_1^3c_2^2c_3^5 - 48ah_1^2c_1^2c_2^3c_3^5 \\
& -60ah_1^2c_1^2c_2^4c_3^4 - 60ah_1^4c_1^2c_2^4c_3^2 - 20ah_1^11c_2 - 3ah_1^9c_1^3 - 15ah_1^11c_1 \\
& +40ah_1^10c_3^2 - ah_1^7c_3^5 + 10ah_1^8c_3^4 + 12ah_1^10c_1^2 + 24ah_1^10c_2^2 - 2ah_1^4c_1^3c_3^5 \\
& -8ah_1^4c_2^3c_3^5 - 30ah_1^7c_1^2c_3^3 + 60ah_1^8c_1^2c_3^2 + 3ah_1^3c_2^4c_3^5 - 5ah_1^7c_2^4c_3 \\
& -40ah_1^7c_2^3c_3^2 - 10ah_1^7c_1^3c_3^2 - 20ah_1^7c_2c_3^4 - 15ah_1^7c_1c_3^4 - 6ah_1^7c_1^3c_2^2
\end{aligned}
$$

$$
\begin{aligned}
&-3ah_1^7c_1c_2^4 - 12ah_1^7c_1^2c_2^3 - 60ah_1^7c_2^2c_3^3 + 24ah_1^8c_1c_2^3 + 8ah_1^8c_1^3c_2 \\
&+36ah_1^8c_1^2c_2^2 + 80ah_1^8c_2c_3^3 + 120ah_1^8c_2^2c_3^2 + 60ah_1^8c_1c_3^3 + 40ah_1^8c_2^3c_3 \\
&+10ah_1^8c_1^3c_3 + 80ah_1^10c_2c_3 + 60ah_1^10c_1c_3 + 48ah_1^10c_1c_2 - 45ah_1^9c_1^2c_3 \\
&-90ah_1^9c_2^2c_3 - 90ah_1^9c_1c_3^2 - 120ah_1^9c_2c_3^2 - 54ah_1^9c_1c_2^2 - 36ah_1^9c_1^2c_2 \\
&+3ah_1^5c_1^2c_3^5 + 6ah_1^5c_2^2c_3^5 + 5ah_1^5c_1^3c_3^4 + 10ah_1^5c_2^4c_3^3 + 20ah_1^5c_2^3c_3^4 \\
&-20ah_1^7c_1^3c_2c_3 - 40ah_1c_1^3h_2c_2^3c_3^4 - 480ah_1^3h_2c_1^2c_2^3c_3^3 + 135ah_1^8h_2c_1^2c_3 \\
&+270ah_1^8h_2c_2^2c_3 + 360ah_1^8h_2c_2c_3^2 + 270ah_1^8h_2c_1c_3^2 + 540ah_1^8h_2c_1c_2c_3 \\
&-12ah_1^11h_2 + 600ah_1^4c_1h_2c_2^3c_3^3 + 900ah_1^4h_2c_1^2c_2^2c_3^3 + 36ah_1^2c_1h_2c_2^3c_3^5 \\
&+54ah_1^2h_2c_1^2c_2^2c_3^5 + 140ah_1^6h_2c_2c_3^4 - 6ah_1h_2c_1c_2^4c_3^5 + 4ac_1^3h_2c_2^3c_3^5 \\
&+45ah_1^2h_2c_1c_2^4c_3^4 - 120ah_1^9h_2c_1c_2 - 200ah_1^9h_2c_2c_3 - 150ah_1^9h_2c_1c_3 \\
&-60ah_1^9h_2c_2^2 + 280ah_1^6h_2c_2^3c_3^2 + 420ah_1^6h_2c_2^2c_3^3 + 210ah_1^6h_2c_1^2c_3^3 \\
&-480ah_1^7h_2c_2^2c_3^2 - 240ah_1^7h_2c_1^2c_3^2 - 40ah_1^7c_1^3h_2c_3 - 320ah_1^7h_2c_2c_3^3 \\
&-32ah_1^7c_1^3h_2c_2 - 160ah_1^7h_2c_2^3c_3 - 720ah_1^7h_2c_1c_2^2c_3 - 480ah_1^7h_2c_1^2c_2c_3 \\
&+3ah_2c_1^2c_2^4c_3^5 - 30ah_1^9h_2c_1^2 - 100ah_1^9h_2c_3^2 + 44ah_1^10h_2c_2 + 33ah_1^10h_2c_1 \\
&+55ah_1^10h_2c_3 + 3ah_1^2h_2c_2^4c_3^5 - 144ah_1^7h_2c_1^2c_2^2 + 5ah_1^4c_1^3c_2^4h_2 \\
&+50ah_1^4h_2c_2^4c_3^3 - 48ah_1^3h_2c_1^2c_2c_3^5 - 72ah_1^3h_2c_1c_2^2c_3^5 \\
&-240ah_1^3c_1^3h_2c_2^2c_3^3 - 120ah_1^3h_2c_1^2c_2^4c_3^2 - 120ah_1^3h_2c_1c_2^4c_3^3 \\
&+630ah_1^6h_2c_1^2c_2^2c_3 - 160ah_1^3c_1^3h_2c_2^3c_3^2 - 20ah_1^3c_1^3c_2^4h_2c_3 \\
&-30ah_1h_2c_1^2c_2^4c_3^4 - 24ah_1h_2c_1^2c_2^3c_3^5 - 12ah_1c_1^3h_2c_2^2c_3^5 \\
&-20ah_1c_1^3c_2^4h_2c_3^3 + 7ah_1^6h_2c_3^5 + 108ah_1^8h_2c_1^2c_2 + 162ah_1^8h_2c_1c_2^2 \\
&+36ah_1^8h_2c_2^3 + 9ah_1^8c_1^3h_2 + 90ah_1^8h_2c_3^3 + 5ac_1^3c_2^4h_2c_3^4 \\
&+840ah_1^6h_2c_1c_2c_3^3 - 18ah_1^5h_2c_1^2c_2^4 - 24ah_1^5c_1^3h_2c_2^3 - 16ah_1^3h_2c_2^3c_3^5 \\
&-18ah_1^5h_2c_1c_3^5 - 720ah_1^5h_2c_1^2c_2c_3^3 - 1080ah_1^5h_2c_1c_2^2c_3^3 \\
&+12ah_1^2c_1^3h_2c_2c_3^5 + 180ah_1^2h_2c_1^2c_2^3c_3^4 + 840ah_1^6h_2c_1^2c_2c_3^2 \\
&+70ah_1^6c_1^3h_2c_3^2 + 120ah_1^2c_1^3c_2^3h_2c_3^3 - 240ah_1^5h_2c_2^3c_3^3 \\
&-360ah_1^5h_2c_1^2c_2^3c_3 - 90ah_1^5h_2c_1c_2^4c_3 - 240ah_1^5c_1^3h_2c_2c_3^2 \\
&-720ah_1^5c_1h_2c_2^3c_3^2 - 1080ah_1^5h_2c_1^2c_2^2c_3^2 - 90ah_1^5h_2c_1^2c_3^4 \\
&-60ah_1^5h_2c_2^4c_3^2 - 360ah_1^5h_2c_1c_2c_3^4 - 24ah_1^5h_2c_2c_3^5 - 180ah_1^5h_2c_2^2c_3^4 \\
&-180ah_1^5c_1^3h_2c_2^2c_3 - 60ah_1^5c_1^3h_2c_3^3 + 200ah_1^4c_1^3h_2c_2c_3^3 \\
&+15ah_1^4h_2c_1^2c_3^5 + 30ah_1^4h_2c_2^2c_3^5 + 25ah_1^4c_1^3h_2c_3^4 + 100ah_1^4c_2^3h_2c_3^4 \\
&+600ah_1^4c_1^2c_2^3h_2c_3^2 + 150ah_1^4c_1c_2^4h_2c_3^2 + 300ah_1^4c_1^3c_2^2h_2c_3^2 \\
&+75ah_1^4h_2c_1^2c_2^4c_3 + 450ah_1^4c_1c_2^2h_2c_3^4 + 42ah_1^6c_1^3h_2c_2^2 \\
&-360ah_1^3c_1^2c_2^2h_2c_3^4 - 20ah_1^3c_2^4h_2c_3^4 - 4ah_1^3c_1^3h_2c_3^5 \\
&-80ah_1^3c_1^3c_2h_2c_3^4 - 240ah_1^3c_1c_2^3h_2c_3^4 + 300ah_1^4c_1^2c_2h_2c_3^4 \\
&+100ah_1^4c_1^3h_2c_2^3c_3 + 60ah_1^4h_2c_1c_2c_3^5 + 1260ah_1^6h_2c_1c_2^2c_3^2
\end{aligned}
$$

$$+30ah_1^2c_1^3c_2^4h_2c_3^2 + 90ah_1^2c_1^2c_2^4h_2c_3^3 + 90ah_1^2c_1^3h_2c_2^2c_3^4$$
$$+84ah_1^6h_2c_1^2c_2^3 + 21ah_1^6h_2c_1c_2^4 + 35ah_1^6h_2c_2^4c_3 - 96ah_1^7c_1h_2c_2^3$$
$$-960ah_1^7h_2c_1c_2c_3^2 - 240ah_1^7h_2c_1c_3^3 + 420ah_1^6c_1h_2c_2^3c_3$$
$$+140ah_1^6c_1^3h_2c_2c_3 + 105ah_1^6h_2c_1c_3^4 - 40ah_1^7h_2c_3^4,$$

where a, b and c are rational functions, and $c_k \in \mathbb{C}$ for $k = 1, \ldots, 4$. The corresponding B_k's are:

$$B_1 = 1, \tag{A.7}$$
$$B_2 = h_1'(h_2 - h_1), \tag{A.8}$$
$$B_3 = h_1'(h_2 - h_1), \tag{A.9}$$
$$B_4 = 3(h_1')^2(h_2 - h_1), \tag{A.10}$$
$$B_5 = 4(h_1')^3(h_2 - h_1), \tag{A.11}$$
$$B_6 = 6(h_1')^5(h_2 - h_1). \tag{A.12}$$

In the above expressions

$$h_j = \frac{\beta - \delta g_j}{\gamma g_j - \alpha}.$$

Note that a, b, c and c_k correspond to $a(z), b(z), c(z)$ and τ_k in Theorem 3.2.1. Note also that here $g_2 \neq \alpha/\gamma$.

Appendix B: The coefficients in Theorem 3.2.3

Also the coefficient functions in Theorem 3.2.3 are very complicated, and so we present them here. They were also calculated by Maple© mathematical software. The coefficients C_k, $k = 1, \ldots, 5$ are:

$$\begin{aligned} C_1 = {} & -4ah_2h_1c_1c_2 + 6ah_2bc_1c_2 - 8ah_2bh_1c_2 - 2ah_2^2bc_1 \\ & -4ab^2h_1h_2 + 3ah_2^2h_1c_2 - ah_2^2c_1c_2 + 3ab^2c_1h_2 + 3ah_2h_1^2c_2 + 3ab^2h_2c_2 \\ & +6ah_2^2bh_1 - 2ah_2^2bc_2 + 3ah_1^2c_1h_2 + 6ah_1^2bh_2 + 3ah_2^2h_1c_1 - 8ah_1bc_1h_2 \\ & -6ah_2^2h_1^2 - ab^2h_2^2 - 2ah_1^2bc_1 - ab^2h_1^2 - ah_1^2c_1c_2 - 2ah_1^2bc_2 + 3ab^2h_1c_2 \\ & -6ab^2c_1c_2 + 3ab^2c_1h_1 + 6ah_1bc_1c_2, \end{aligned} \tag{B.1}$$

$$\begin{aligned} C_2 = {} & -ac_1c_2h_1^2 - ah_1^2c_2c_3 - ac_1h_1^2c_3 - ah_1^2c_3c_4 - ah_1^2c_2c_4 - ac_1h_1^2c_4 \\ & -6ah_2^2h_1^2 + 3ac_1c_2h_1c_4 + 3ac_1c_2c_3h_1 - 6ac_1c_2c_3c_4 + 3ah_1c_2c_3c_4 \\ & +3ac_1h_1c_3c_4 + 3ac_1c_2c_3h_2 - 4ac_1c_2h_2h_1 + 3ac_1c_2h_2c_4 - 4ah_2h_1c_3c_4 \\ & -4ah_2c_2h_1c_4 - 4ah_2c_2c_3h_1 + 3ah_2c_2c_3c_4 - 4ac_1h_2h_1c_4 - 4ac_1h_2c_3h_1 \\ & +3ac_1h_2c_3c_4 + 3ah_2h_1^2c_3 + 3ah_2c_2h_1^2 + 3ac_1h_2h_1^2 + 3ah_2h_1^2c_4 \\ & -ac_1c_2h_2^2 + 3ah_2^2h_1c_3 + 3ah_2^2c_2h_1 - ah_2^2c_2c_3 + 3ac_1h_2^2h_1 - ac_1h_2^2c_3 \\ & -ah_2^2c_2c_4 - ac_1h_2^2c_4 + 3ah_2^2h_1c_4 - ah_2^2c_3c_4, \end{aligned} \tag{B.2}$$

$$\begin{aligned}
C_3 = & -2ah_1^3c_1c_3^2 + 4ac_1^2h_1^2c_3^2 + 4ac_1^2c_2^2h_1^2 - 2ah_1^3c_2^2c_3 + 4ah_1^2c_2^2c_3^2 \\
& -2ac_1^2h_1^3c_3 - 2ac_1^2h_1^3c_2 - 2ah_1^3c_1c_2^2 + 20ac_1^2c_2^2c_3^2 - 2ah_1^3c_2c_3^2 \\
& -20ac_1^2c_2^2h_1c_3 - 8ah_1^3c_1c_2c_3 + 16ah_1^2c_1c_2c_3^2 + 16ah_1^2c_1c_2^2c_3 \\
& -20ah_1c_1c_2^2c_3^2 + 16ac_1^2h_1^2c_2c_3 - 20ac_1^2h_1c_2c_3^2 - 20ah_2^3h_1^2c_2 \\
& -20ah_2^2h_1^3c_2 + 4ah_2^3h_1c_2^2 + 12ah_2^2h_1^2c_2^2 + 4ah_2h_1^3c_2^2 + 16ah_2^3c_1h_1c_3 \\
& -18ah_2h_1^2c_2c_3^2 + 4ac_1^2h_2^3h_1 + 16ah_2^3c_1c_2h_1 - 8ah_2^3c_1c_2c_3 + 48ah_2^2c_1c_2h_1^2 \\
& +48ah_2^2c_1h_1^2c_3 - 18ah_2^2c_1c_2^2h_1 + 16ah_2^2c_1c_2^2c_3 - 18ah_2c_1c_2^2h_1^2 \\
& -18ah_2^2h_1c_2^2c_3 - 2ac_1^2h_2^3c_2 - 2ac_1^2h_2^3c_3 + 12ac_1^2h_1^2h_2^2 + 4ac_1^2h_1^3h_2 \\
& +4ac_1^2c_2^2h_2^2 - 20ah_2^3c_1h_1^2 - 20ah_2^2c_1h_1^3 - 2ah_2^3c_1c_2^2 + 4ac_1^2h_2^2c_3^2 \\
& +16ah_2h_1^3c_2c_3 + 16ah_1^3c_1h_2c_3 - 18ac_1^2h_1^2h_2c_3 + 12ac_1^2c_2^2h_2h_1 \\
& -20ac_1^2c_2^2h_2c_3 - 18ac_1^2h_2^2h_1c_3 - 18ac_1^2h_2^2c_2h_1 + 16ac_1^2h_2^2c_2c_3 \\
& -18ac_1^2h_2c_2h_1^2 + 16ah_2c_1h_1^3c_2 - 72ah_1^2c_1c_2h_2c_3 - 18ah_2^2h_1c_2c_3^2 \\
& +16ah_2^3h_1c_2c_3 + 48ah_2^2h_1^2c_2c_3 - 20ac_1^2h_2c_2c_3^2 + 48ah_2c_1h_1c_2c_3^2 \\
& +4ah_2^3h_1c_3^2 - 2ah_2^3c_2c_3^2 + 12ah_2^2h_1^2c_3^2 + 4ah_2^2c_2^2c_3^2 + 4ah_2h_1^3c_3^2 \\
& -2ah_2^3c_1c_3^2 - 18ah_2^2c_1h_1c_3^2 - 72ah_2^2c_1c_2h_1c_3 + 16ah_2^2c_1c_2c_3^2 \\
& -18ah_2c_1h_1^2c_3^2 + 48ah_2c_1c_2^2h_1c_3 - 20ah_2c_1c_2^2c_3^2 + 12ac_1^2h_2h_1c_3^2 \\
& -18ah_2h_1^2c_2^2c_3 + 12ah_2h_1c_2^2c_3^2 + 20ah_2^3h_1^3 + 48ac_1^2h_1c_2h_2c_3 \\
& -2ah_2^3c_2^2c_3 - 20ah_2^2h_1^3c_3 - 20ah_2^3h_1^2c_3,
\end{aligned} \tag{B.3}$$

$$\begin{aligned}
C_4 = & -105ah_2^4c_2h_1^3 - 15ah_2^3c_2^2c_3^3 - 5ah_2h_1^4c_3^3 - 30ah_2^2c_2^3h_1^3 \\
& -5ah_2^4c_2^3h_1 - 30ah_2^3c_2^3h_1^2 - 30ah_2^2h_1^3c_3^3 + 15ah_2^2c_2^3c_3^3 - 5ah_2^4h_1c_3^3 \\
& +3ah_2^4c_2c_3^3 - 30ah_2^3h_1^2c_3^3 + 120ah_2^3c_2^2h_1^3 + 45ah_2^4c_2^2h_1^2 + 48ac_1^2h_2h_1^3c_3^2 \\
& -180ah_2^2c_1h_1^3c_2^2 - 30ac_1^2h_2h_1^2c_3^3 + 360ac_1^2h_2c_2^2h_1c_3^2 - 105ac_1^2h_2c_2^2c_3^3 \\
& -30ah_2^4c_1h_1c_2^2 + 288ah_2c_1h_1^3c_2^2c_3 - 30ah_2c_1h_1^4c_2^2 - 180ah_2^3c_1h_1^2c_2^2 \\
& +90ah_2^2c_1h_1^4c_2 + 90ah_2^4c_1h_1^2c_2 + 240ah_1c_1c_2^3h_2c_3^2 + 135ah_2^4c_2h_1^2c_3 \\
& -180ah_1c_1c_2^3h_2^2c_3 + 360ah_2^3h_1^3c_2c_3 + 648ah_2^2c_1h_1^2c_2c_3^2 \\
& -90ac_1^2h_2^3h_1^2c_3 + 144ac_1^2h_2^3c_2h_1c_3 - 90ac_1^2h_2^2h_1^3c_3 \\
& -270ac_1^2h_2^2c_2^2h_1c_3 - 90ah_2^2c_2^3h_1c_3^2 - 45ah_2^4c_2h_1c_3^2 \\
& +240ah_2^3c_1h_1^3c_2 + 108ah_2^2c_2^3h_1^2c_3 - 270ah_2^3c_2^2h_1^2c_3 \\
& -180ah_2^2c_1h_1c_2c_3^3 - 180ah_2c_1h_1^2c_2c_3^3 + 144ah_2^3c_2^2h_1c_3^2 \\
& -270ac_1^2h_2^2c_2h_1c_3^2 + 45ac_1^2h_2^2c_2c_3^3 + 324ac_1^2h_2^2c_2h_1^2c_3 \\
& -90ac_1^2h_2^2c_2h_1^3 - 15ac_1^2h_2^4c_2h_1 - 90ac_1^2h_2^3c_2h_1^2 - 15ac_1^2h_2h_1^4c_3 \\
& -270ac_1^2h_2c_2^2h_1^2c_3 + 48ac_1^2h_2c_2^2h_1^3 + 48ac_1^2h_2^3c_2^2h_1 \\
& +108ac_1^2h_2^2c_2^2h_1^2 + 40ac_1^2h_2^3h_1^3 + 15ac_1^2h_2^2h_1^4 + 15ac_1^2h_2^4h_1^2 \\
& -30ac_1^2c_2^3h_2^2h_1 + 32ah_2^3c_1h_1c_3^3 + 2ah_2^4c_1c_3^3 + 9ah_2^4c_2^2c_3^2 \\
& +32ah_2c_1h_1^3c_3^3 - 70ah_2c_1c_2^3c_3^3 + 48ah_2^3h_1c_2c_3^3 + 108ah_2^2h_1^2c_2c_3^3
\end{aligned} \tag{B.4}$$

$$
\begin{aligned}
&-90ah_2^2h_1c_2^2c_3^3+48ah_2h_1^3c_2c_3^3-90ah_2h_1^2c_2^2c_3^3-70ah_2^4c_1h_1^3\\
&-70ah_2^3c_1h_1^4+2ah_2^4c_1c_2^3-5ac_1^2c_2^3h_2^3+144ac_1^2h_1^3c_2h_2c_3\\
&-180ah_1^2c_1c_2^3h_2c_3+120ac_1^2h_2h_1c_2c_3^3-270ac_1^2h_1^2c_2h_2c_3^2\\
&-105ah_2^3h_1^4c_3+144ah_1^3c_2^2h_2c_3^2-90ah_1^2c_2^3h_2c_3^2-30ah_1^4c_1h_2c_3^2\\
&+45ac_1^2c_2^3h_2^2c_3-105ac_1^2c_2^3h_2c_3^2-90ah_2^4c_1h_1c_2c_3\\
&+288ah_2^3c_1h_1c_2c_3^2-105ah_2^4h_1^3c_3+45ah_2^2h_1^4c_3^2+48ac_1^2h_2^3h_1c_3^2\\
&+9ac_1^2h_2^4c_2c_3-45ac_1^2h_2^3c_2c_3^2+108ac_1^2h_2^2h_1^2c_3^2-45ac_1^2h_2^3c_2^2c_3\\
&-15ac_1^2h_2^4h_1c_3-15ac_1^2h_1^4c_2h_2+32ah_1c_1c_2^3h_2^3+72ah_1^2c_1c_2^3h_2^2\\
&+32ah_1^3c_1c_2^3h_2+40ah_2h_1c_2^3c_3^3-180ah_2^2c_1h_1^3c_3^2-30ah_2^3c_1c_2^3c_3\\
&+90ah_2^2c_1c_2^3c_3^2-30ah_2^4c_1h_1c_3^2+18ah_2^4c_1c_2c_3^2-180ah_2^3c_1h_1^2c_3^2\\
&+18ah_2^4c_1c_2^2c_3-90ah_2^3c_1c_2^2c_3^2+135ah_2^2h_1^4c_2c_3+48ah_2^3c_2^3h_1c_3\\
&-540ah_2^3c_1h_1^2c_2c_3-540ah_2^2c_1h_1^3c_2c_3+648ah_2^2c_1h_1^2c_2^2c_3\\
&+135ac_1^2h_2^2c_2^2c_3^2-270ah_2^2h_1^3c_2^2c_3-540ah_2c_1h_1^2c_2^2c_3^2\\
&-30ac_1^2h_2^2h_1c_3^3+15ah_1^2c_2^3c_3^3-15ah_1^3c_2^3c_3^2-15ah_1^3c_2^2c_3^3\\
&+135ac_1^2h_1^2c_2^2c_3^2-105ac_1^2h_1c_2^2c_3^3+18ah_1^4c_1c_2^2c_3-5ac_1^2c_2^3h_1^3\\
&+3ac_1^2h_1^4c_2^2-90ah_1^3c_1c_2^2c_3^2+90ah_1^2c_1c_2^2c_3^3+90ah_1^2c_1c_2^3c_3^2\\
&-70ah_1c_1c_2^3c_3^3-45ac_1^2h_1^3c_2c_3^2+45ac_1^2h_1^2c_2c_3^3-30ah_1^3c_1c_2^3c_3\\
&+9ac_1^2h_1^4c_2c_3-45ac_1^2h_1^3c_2^2c_3+45ac_1^2c_2^3h_1^2c_3-105ac_1^2c_2^3h_1c_3^2\\
&+2ah_1^4c_1c_2^3-5ac_1^2h_1^3c_3^3+9ah_1^4c_2^2c_3^2+3ac_1^2h_1^4c_3^2+2ah_1^4c_1c_3^3\\
&+3ah_1^4c_2^3c_3+3ah_1^4c_2c_3^3+70ac_1^2c_2^3c_3^3+18ah_1^4c_1c_2c_3^2\\
&-30ah_1^3c_1c_2c_3^3+240ah_2^3c_1h_1^3c_3-45ah_1^4c_2^2h_2c_3+48ah_1^3c_2^3h_2c_3\\
&+90ah_2^2c_1h_1^4c_3+90ah_2^4c_1h_1^2c_3+120ac_1^2c_2^3h_2h_1c_3-270ah_2^3h_1^2c_2c_3^2\\
&-270ah_2^2h_1^3c_2c_3^2+324ah_2^2h_1^2c_2^2c_3^2-5ac_1^2h_2^3c_3^3-105ah_2^3h_1^4c_2\\
&+45ah_2^2h_1^4c_2^2-5ah_2h_1^4c_2^3-45ah_1^4c_2h_2c_3^2+288ah_2^3c_1h_1c_2^2c_3\\
&-540ah_2^2c_1h_1c_2^2c_3^2+288ah_1^3c_1c_2h_2c_3^2-30ac_1^2c_2^3h_2h_1^2\\
&+3ac_1^2h_2^4c_2^2+3ac_1^2h_2^4c_3^2+240ah_2c_1h_1c_2^2c_3^3-90ah_1^4c_1c_2h_2c_3\\
&-45ah_2^4c_2^2h_1c_3+120ah_2^3h_1^3c_3^2+3ah_2^4c_2^3c_3-15ah_2^3c_2^3c_3^2\\
&+70ah_2^4h_1^4-30ah_2^3c_1c_2c_3^3+72ah_2^2c_1h_1^2c_3^3+90ah_2^2c_1c_2^2c_3^3\\
&+45ah_2^4h_1^2c_3^2,
\end{aligned}
$$

$$
\begin{aligned}
C_5 = {} & 2520ac_1^3h_1^3c_2^2c_3^4-4ac_1^3h_1^6c_2^3-3ah_1^6c_1c_3^5+84ah_1^3c_2^4c_3^5\\
&-15ah_1^6c_1^2c_3^4-40ah_1^6c_2^3c_3^3-140ah_1^4c_2^4c_3^4+2310ac_1^3c_2^4h_1c_3^4\\
&-924ah_2^6h_1^6-10ac_1^3h_1^6c_3^3+42ah_1^5c_2^2c_3^5+140ah_1^5c_2^3c_3^4\\
&-924ac_1^3c_2^4c_3^5+21ah_1^5c_1^2c_3^5+7ac_1^3c_2^4h_1^5+35ac_1^3h_1^5c_3^4\\
&-4ah_1^6c_2c_3^5-10ah_1^6c_2^4c_3^2+70ah_1^5c_2^4c_3^3-112ah_1^4c_2^3c_3^5\\
&-30ah_1^6c_2^2c_3^4+84ah_1^5c_1c_2c_3^5-630ah_1^2c_1c_2^4c_3^5-336ah_1^4c_1^2c_2c_3^5
\end{aligned}
\qquad \text{(B.5)}
$$

$$
\begin{aligned}
&+1512ah_1^3c_1^2c_2^2c_3^5-2520ah_1^2c_1^2c_2^3c_3^5+1386ah_1c_1^2c_2^4c_3^5\\
&-3ah_1^6c_1^2c_2^4-560ac_1^3h_1^4c_2c_3^4+280ac_1^3h_1^5c_2c_3^3\\
&+420ac_1^3h_1^5c_2^2c_3^2-60ah_1^6c_1c_2c_3^4+210ah_1^5c_1c_2^4c_3^2\\
&+1260ah_1^3c_1c_2^4c_3^4-840ah_1^4c_1c_2^4c_3^3-15ah_1^6c_1c_2^4c_3\\
&+1008ah_1^3c_1c_2^3c_3^5-504ah_1^4c_1c_2^2c_3^5-40ac_1^3h_1^6c_2c_3^2\\
&-2520ah_1^4c_1^2c_2^2c_3^4-28ac_1^3h_1^4c_3^5-2100ac_1^3c_2^4h_1^2c_3^3\\
&-140ac_1^3c_2^4h_1^4c_3-120ah_1^6c_1c_2^3c_3^2-1680ah_1^4c_1c_2^3c_3^4\\
&+840ah_1^5c_1c_2^3c_3^3+630ah_1^5c_1c_2^2c_3^4-180ah_1^6c_1c_2^2c_3^3\\
&+840ac_1^3c_2^4h_1^3c_3^2-1680ac_1^3h_1^4c_2^2c_3^3-30ac_1^3h_1^6c_2^2c_3\\
&-1120ac_1^3h_1^4c_2^3c_3^2-4200ac_1^3h_1^2c_2^3c_3^4+3360ac_1^3h_1^3c_2^3c_3^3\\
&+140ac_1^3h_1^5c_2^3c_3+1260ah_1^5c_1^2c_2^2c_3^3+840ah_1^5c_1^2c_2^3c_3^2\\
&+5040ah_1^3c_1^2c_2^3c_3^4-3360ah_1^4c_1^2c_2^3c_3^3-60ah_1^6c_1^2c_2^3c_3\\
&-840ah_1^4c_1^2c_2^4c_3^2-3150ah_1^2c_1^2c_2^4c_3^4+2520ah_1^3c_1^2c_2^4c_3^3\\
&+105ah_1^5c_1^2c_2^4c_3+336ac_1^3h_1^3c_2c_3^5-1260ac_1^3h_1^2c_2^2c_3^5\\
&+1848ac_1^3h_1c_2^3c_3^5+420ah_1^5c_1^2c_2c_3^4-120ah_1^6c_1^2c_2c_3^3\\
&-180ah_1^6c_1^2c_2^2c_3^2+3150ah_2h_1^4c_1c_2^4c_3^2-6750ah_2^2h_1^4c_2^2c_3^4\\
&-1680ah_2^2h_1^6c_2^2c_3^2+7000ah_2^2h_1^3c_2^3c_3^4+2100ah_2^2h_1^5c_2c_3^4\\
&-336ah_2^2h_1^6c_1c_2^3-12600ah_2^2c_1^2c_2^4h_1^2c_3^2-675ah_2^2c_1^2c_2^4h_1^4\\
&+5670ah_2h_1^2c_1c_2^4c_3^4-2160ah_2h_1^5c_1c_2c_3^4-6720ah_2h_1^3c_1c_2^4c_3^3\\
&+840ah_2h_1^6c_1c_2c_3^3-630ah_2^2c_1c_2^4c_3^5+1750ah_2^2c_1^3h_1^3c_3^4\\
&-2520ah_2^2c_1^2c_2^3c_3^5-675ah_2^2h_1^4c_1c_3^5+1386ah_2c_1^2c_2^4c_3^5\\
&-5040ah_2^2h_1^2c_1^2c_2c_3^5+4200ah_2^2h_1^3c_1c_2c_3^5+378ah_2^2h_1c_2^4c_3^5\\
&-6480ah_2^5c_1^2c_2^2h_1c_3^2-1080ah_2^5c_1^3h_1c_2^2c_3+2100ah_2^5c_1^3h_1^2c_2c_3\\
&+15120ah_2c_1^3h_1^2c_2^3c_3^3+2100ah_2c_1^3h_1^4c_2^3c_3+84ah_2^5c_1c_2c_3^5\\
&+1050ah_2^4c_1c_2^4h_1^3+1512ah_2c_1^3h_1^2c_2c_3^5-3024ah_2c_1^3h_1c_2^2c_3^5\\
&-140ah_2^2h_1^6c_3^4-13500ah_2^2h_1^4c_1c_2c_3^4-8400ah_2^2c_1^3h_1^2c_2c_3^4\\
&-9000ah_2^2c_1^3h_1^4c_2c_3^2-3360ah_2^2h_1^6c_1c_2c_3^2+18900ah_2^2h_1^5c_1c_2^2c_3^2\\
&-26880ah_2h_1^3c_1^2c_2^3c_3^3-4320ah_2h_1^5c_1^2c_2c_3^3+12600ah_2^5h_1^2c_1^2c_2c_3^2\\
&+630ah_2h_1^6c_1^2c_2^2c_3+9450ah_2h_1^4c_1c_2^2c_3^4-1120ah_2^5c_1^3h_1^3c_3\\
&+4536ah_2^5h_1^4c_1^2c_2+6804ah_2^5h_1^4c_1c_2^2-6048ah_2^5h_1^5c_1c_2\\
&-2688ah_2^5h_1^3c_1c_2^3-1440ah_2^5c_1^3c_2h_1c_3^2-2160ah_2^5h_1c_1^2c_2^3c_3\\
&-13440ah_2c_1^3h_1^3c_2^2c_3^3+315ah_2^5h_1^2c_1c_2^4+1260ah_2^5h_1^2c_1^2c_2^3\\
&+9450ah_2^5h_1^2c_1^2c_2^2c_3+6804ah_2h_1^2c_1^2c_2^2c_3^5+420ah_2h_1^6c_1c_2^3c_3\\
&+12600ah_2h_1^4c_1c_2^3c_3^3-2100ah_2^4h_1^6c_3^2-5040ah_2c_1^3c_2^4h_1c_3^3\\
&-6480ah_2h_1^5c_1c_2^2c_3^3+420ah_2h_1^6c_2^2c_3^3+6300ah_2^5h_1^2c_1c_2^3c_3
\end{aligned}
$$

$$
\begin{aligned}
&+5670ah_2^5h_1^4c_1^2c_3 + 35ah_2h_1^6c_2^4c_3 - 1120ah_2c_1^3c_2^4h_1^3c_3 \\
&+315ah_2h_1^4c_1^2c_3^5 - 224ah_2c_1^3h_1^3c_3^5 - 108ah_2h_1^5c_1c_3^5 \\
&-360ah_2c_1^3h_1^5c_3^3 + 1050ah_2h_1^4c_2^4c_3^3 + 210ah_2h_1^6c_1^2c_3^3 \\
&-504ah_2^2h_1^6c_1^2c_2^2 + 1050ah_2^4h_1c_2^4c_3^3 + 84ah_2^3c_1^3h_1^6 \\
&-630ah_2^6h_1^4c_1^2 - 140ah_2^4c_1^3c_2^4c_3 - 10080ah_2^5h_1^5c_2c_3 \\
&+3150ah_2^5h_1^2c_1^2c_3^3 - 840ah_2^6h_1^2c_1c_3^3 + 2520ah_2^6h_1^3c_1c_3^2 \\
&+2520ah_2^6h_1^3c_2^2c_3 + 840ah_2^6h_1c_1c_2c_3^3 - 3360ah_2^6h_1^2c_1c_2c_3^2 \\
&+2100ah_2^2c_1^3h_1^5c_2c_3 - 180ah_2^6c_1c_2^2c_3^3 - 13440ah_2^5h_1^3c_1^2c_2c_3 \\
&-6720ah_2^5h_1^3c_1c_3^3 + 11340ah_2^5h_1^4c_1c_3^2 - 560ah_2^6h_1^2c_2^3c_3 \\
&+22680ah_2^5h_1^4c_1c_2c_3 - 900ah_2^2c_1^3c_2^3h_1^4 - 9000ah_2^2h_1^4c_2^3c_3^3 \\
&-40500ah_2^2h_1^4c_1^2c_2^2c_3^2 + 14000ah_2^2c_1^3h_1^3c_2c_3^3 + 1512ah_2^2c_1^3h_1c_2c_3^5 \\
&-2100ah_2^2c_1^3c_2^4c_3^3 - 27000ah_2^2h_1^4c_1c_2^3c_3^2 - 6480ah_2h_1^5c_1^2c_2^2c_3^2 \\
&-20160ah_2h_1^3c_1^2c_2^2c_3^4 + 840ah_2h_1^6c_1^2c_2c_3^2 + 6300ah_2h_1^4c_1^2c_2c_3^4 \\
&-20160ah_2^5h_1^3c_1c_2^2c_3 + 4536ah_2^2h_1c_1c_2^3c_3^5 + 6804ah_2^2h_1c_1^2c_2^2c_3^5 \\
&+22680ah_2h_1^2c_1^2c_2^3c_3^4 - 7560ah_2h_1c_1^2c_2^4c_3^4 + 4200ah_2^3h_1^2c_1c_2c_3^5 \\
&+840ah_2^6c_1^2c_2h_1c_3^2 - 2520ah_2^6h_1^4c_1c_2 - 2688ah_2^3h_1c_1^2c_2c_3^5 \\
&-7560ah_2^2h_1^2c_1c_2^2c_3^5 + 840ah_2^3h_1^6c_3^3 - 1120ah_2^3h_1^5c_3^4 + 84ah_2^3c_2^4c_3^5 \\
&+140ah_2^5c_1^3c_2^3c_3 + 22680ah_2^2h_1c_1^2c_2^3c_3^4 + 11340ah_2^4h_1^5c_2^2c_3 \\
&-13500ah_2^2h_1^4c_1^2c_2^3c_3 + 1260ah_2^2h_1^5c_1^2c_2^3 - 3375ah_2^2h_1^4c_1^2c_3^4 \\
&-840ah_2^2h_1^6c_1^2c_3^2 + 5040ah_2^6h_1^3c_1c_2c_3 - 60ah_2^6c_1^2c_2^3c_3 \\
&-3150ah_2^6h_1^4c_1c_3 - 1120ah_2^6h_1^2c_2c_3^3 - 40ah_2^6c_1^3c_2c_3^2 \\
&+378ah_2^5c_1^3h_1^4 - 5040ah_2^5h_1^5c_3^2 - 28ah_2^6h_1^2c_2^4 - 10ah_2^6c_2^4c_3^2 \\
&-4ah_2^6c_2c_3^5 - 40ah_2^6c_2^3c_3^3 + 70ah_2^5c_2^4c_3^3 + 1848ah_2^5h_1^6c_2 \\
&-560ah_2^2h_1^6c_2^3c_3 + 6300ah_2^2h_1^5c_2^2c_3^3 - 1120ah_2^2h_1^6c_2c_3^3 \\
&+7000ah_2^2c_1^3c_2^3h_1^3c_3 + 15120ah_2^2c_1^3c_2^3h_1c_3^3 - 1080ah_2^5c_2^2h_1c_3^4 \\
&+1050ah_2^2c_1^3h_1^5c_3^2 + 1050ah_2^5c_1^3h_1^2c_3^2 - 1440ah_2^5h_1c_2^3c_3^3 \\
&-540ah_2^5c_1^2h_1c_3^4 + 525ah_2^5h_1^2c_2^4c_3 + 4200ah_2^5h_1^2c_2^3c_3^2 \\
&-26880ah_2^5h_1^3c_1c_2c_3^2 - 144ah_2^5h_1c_2c_3^5 - 4320ah_2^5h_1c_1^2c_2c_3^3 \\
&+4200ah_2c_1^3h_1^4c_2c_3^3 + 140ah_2c_1^3h_1^6c_2c_3 - 2160ah_2h_1^5c_1^2c_2^3c_3 \\
&+11340ah_2h_1^2c_1^2c_2^4c_3^3 + 1575ah_2h_1^4c_1^2c_2^4c_3 + 18900ah_2h_1^4c_1^2c_2^2c_3^3 \\
&+2100ah_2h_1^4c_2^3c_3^4 + 105ah_2c_1^3c_2^4h_1^4 + 140ah_2h_1^6c_2c_3^4 \\
&-6480ah_2^5h_1c_1c_2^2c_3^3 + 18900ah_2^5h_1^2c_1c_2^2c_3^2 + 2310ah_2c_1^3c_2^4c_3^4 \\
&-360ah_2h_1^5c_2^4c_3^2 - 540ah_2h_1^5c_1^2c_3^4 + 525ah_2c_1^3h_1^4c_3^4 + 70ah_2c_1^3h_1^6c_3^2 \\
&+105ah_2h_1^6c_1c_3^4 + 280ah_2h_1^6c_2^3c_3^2 - 1120ah_2h_1^3c_2^4c_3^4 + 420ah_2^5c_1^3c_2^2c_3^2 \\
&+1260ah_2h_1^6c_1c_2^2c_3^2 + 3780ah_2c_1^3c_2^4h_1^2c_3^2 + 11340ah_2^2c_1^3h_1c_2^2c_3^4
\end{aligned}
$$

$$
\begin{aligned}
&+21000ah_2^2c_1^3h_1^3c_2^2c_3^2 + 11340ah_2c_1^3h_1^2c_2^2c_3^4 + 6300ah_2c_1^3h_1^4c_2^2c_3^2\\
&-1440ah_2c_1^3h_1^5c_2c_3^2 - 4480ah_2c_1^3h_1^3c_2c_3^4 - 1680ah_2^4c_1c_2^3c_3^4\\
&-2520ah_2^4c_1^2c_2^2c_3^4 - 1120ah_2^4c_1^3c_2^3c_3^2 - 840ah_2^4c_1^2c_2^4c_3^2\\
&-560ah_2^4c_1^3c_2c_3^4 + 1575ah_2^4h_1c_1^2c_2^4c_3 + 2100ah_2^4c_1^3h_1c_2^3c_3\\
&+105ah_2^4c_1^3c_2^4h_1 + 1050ah_2^3h_1^2c_1^2c_3^5 - 1600ah_2^3h_1^3c_2c_3^5\\
&+2100ah_2^3h_1^2c_2^2c_3^5 - 900ah_2^4c_1^3h_1^2c_2^3 + 9450ah_2^4c_1c_2^2h_1c_3^4\\
&+6300ah_2^4c_1^2c_2h_1c_3^4 - 37800ah_2^4h_1^4c_1c_2^2c_3 - 675ah_2^4h_1^2c_1^2c_2^4\\
&-13500ah_2^4h_1^2c_1c_2c_3^4 + 1260ah_2^4h_1c_1c_2c_3^5 - 504ah_2^4c_1c_2^2c_3^5\\
&-336ah_2^4c_1^2c_2c_3^5 + 1512ah_2^6h_1^3c_1c_2^2 - 120ah_2^6c_1^2c_2c_3^3\\
&-3375ah_2^4c_1c_2^4h_1^2c_3 - 13500ah_2^4c_1^2c_2^3h_1^2c_3 + 525ah_2^4c_1^3h_1c_3^4\\
&+7000ah_2^4h_1^3c_2c_3^4 - 6750ah_2^4h_1^2c_2^2c_3^4 + 5250ah_2^4h_1^3c_1c_3^4\\
&-3375ah_2^4h_1^2c_1^2c_3^4 - 540ah_2h_1^5c_1c_2^4c_3 - 8960ah_2c_1^3h_1^3c_2^3c_3^2\\
&-10080ah_2c_1^3h_1c_2^3c_3^4 - 504ah_2^6h_1^2c_1^2c_2^2 + 420ah_2^6h_1c_1c_2^3c_3\\
&-4200ah_2^6h_1^4c_2c_3 - 112ah_2^6c_1^3h_1^2c_2 - 540ah_2^5h_1c_1c_2^4c_3\\
&+1260ah_2^6c_1c_2^2h_1c_3^2 - 336ah_2^6h_1^2c_1c_2^3 + 21ah_2^5c_1^2c_3^5\\
&+630ah_2^6h_1c_1^2c_2^2c_3 + 280ah_2^6c_2^3h_1c_3^2 + 140ah_2^6c_1^3h_1c_2c_3\\
&+210ah_2^6h_1c_1^2c_3^3 - 2520ah_2^6h_1^2c_1c_2^2c_3 - 6720ah_2^5h_1^3c_1^2c_3^2\\
&-120ah_2^6c_1c_2^3c_3^2 - 180ah_2^6c_1^2c_2^2c_3^2 - 140ah_2^6c_1^3h_1^2c_3\\
&+21ah_2^6h_1c_1c_2^4 + 84ah_2^6h_1c_1^2c_2^3 + 70ah_2^6c_1^3h_1c_3^2\\
&-1680ah_2^6h_1^2c_1^2c_2c_3 + 420ah_2^6h_1c_2^2c_3^3 - 30ah_2^6c_1^3c_2^2c_3\\
&-1680ah_2^6h_1^2c_2^2c_3^2 - 60ah_2^6c_1c_2c_3^4 - 1260ah_2^6h_1^4c_2^2 + 105ah_2^2h_1^5c_3^5\\
&-1120ah_2^5h_1^3c_3^4 - 1440ah_2h_1^5c_2^3c_3^3 + 63000ah_2^2h_1^3c_1^2c_2^2c_3^3\\
&-27000ah_2^2h_1^4c_1^2c_2c_3^3 - 1680ah_2^2h_1^6c_1^2c_2c_3 + 9450ah_2^2h_1^5c_1^2c_2^2c_3\\
&-144ah_2h_1^5c_2c_3^5 + 420ah_2^5c_1^2c_2c_3^4 + 21ah_2h_1^6c_1c_2^4\\
&-40500ah_2^2h_1^4c_1c_2^2c_3^3 - 2520ah_2^2h_1^6c_1c_2^2c_3 + 12600ah_2^2h_1^5c_1c_2c_3^3\\
&-25200ah_2^2c_1^3h_1^2c_2^2c_3^3 - 6750ah_2^2c_1^3h_1^4c_2^2c_3 - 896ah_2h_1^3c_2^3c_3^5\\
&-8960ah_2^5h_1^3c_2c_3^3 + 15120ah_2^5h_1^4c_2c_3^2 - 13440ah_2^5h_1^3c_2^2c_3^2\\
&-15ah_2^6c_1c_2^4c_3 + 140ah_2^6c_2h_1c_3^4 + 35ah_2^6h_1c_2^4c_3 - 15ah_2^6c_1^2c_3^4\\
&+840ah_2^6h_1^3c_3^3 - 30ah_2^6c_2^2c_3^4 + 336ah_2^6h_1^3c_2^3 + 2310ah_2^6h_1^5c_3\\
&-1512ah_2^5h_1^5c_1^2 - 10ah_2^6c_1^3c_3^3 + 1848ah_2^6h_1^5c_2 + 84ah_2^6c_1^3h_1^3\\
&+1386ah_2^5h_1^6c_1 + 140ah_2^5c_2^3c_3^4 + 105ah_2^6c_1h_1c_3^4 + 3360ah_2^6h_1^3c_2c_3^2\\
&+1260ah_2^6h_1^3c_1^2c_3 + 6300ah_2^5h_1^2c_2^2c_3^3 - 2250ah_2^2c_1^3h_1^4c_3^3\\
&-140ah_2^2c_1^3h_1^6c_3 + 42000ah_2^2h_1^3c_1^2c_2^3c_3^2 + 315ah_2^2h_1^5c_1c_2^4\\
&-112ah_2^2c_1^3h_1^6c_2 + 630ah_2^2c_1^3h_1^5c_2^2 + 42000ah_2^2h_1^3c_1c_2^3c_3^3\\
&+6300ah_2^2h_1^5c_1c_2^3c_3 - 12600ah_2^2h_1^2c_1c_2^4c_3^3 - 3375ah_2^2h_1^4c_1c_2^4c_3
\end{aligned}
$$

$$
\begin{aligned}
&-1512ah_2h_1c_1c_2^4c_3^5 + 4536ah_2h_1^2c_1c_2^3c_3^5 + 630ah_2^5c_1^3h_1^2c_2^2\\
&-144ah_2^5c_1^3h_1c_2^3 - 108ah_2^5h_1c_1^2c_2^4 + 42ah_2c_1^3h_1^6c_2^2\\
&-144ah_2c_1^3h_1^5c_2^3 - 6720ah_2h_1^3c_1^2c_2^4c_3^2 + 1575ah_2^5h_1^2c_1c_3^4\\
&-108ah_2^5h_1c_1c_3^5 + 2100ah_2^5h_1^2c_2c_3^4 - 28ah_2^2h_1^6c_2^4\\
&-13440ah_2h_1^3c_1c_2^3c_3^4 - 4320ah_2h_1^5c_1c_2^3c_3^2 - 37800ah_2^2h_1^2c_1^2c_2^2c_3^4\\
&+21000ah_2^2h_1^3c_1^2c_2c_3^4 + 12600ah_2^2h_1^5c_1^2c_2c_3^2 + 5670ah_2^2h_1c_1c_2^4c_3^4\\
&+10500ah_2^2h_1^3c_1c_2^4c_3^2 - 25200ah_2^2h_1^2c_1c_2^3c_3^4 + 3780ah_2^2c_1^3c_2^4h_1c_3^2\\
&+350ah_2^2c_1^3c_2^4h_1^3 - 2100ah_2^2c_1^3c_2^4h_1^2c_3 - 4032ah_2h_1^3c_1c_2^2c_3^5\\
&+1260ah_2h_1^4c_1c_2c_3^5 - 2688ah_2h_1^3c_1^2c_2c_3^5 - 1080ah_2c_1^3h_1^5c_2^2c_3\\
&-224ah_2^5h_1^3c_2^4 + 3500ah_2^4c_1^3h_1^3c_3^2 + 630ah_2^5c_1c_2^2c_3^4\\
&+378ah_2h_1^2c_2^4c_3^5 - 6048ah_2h_1c_1^2c_2^3c_3^5 + 1848ah_2c_1^3c_2^3c_3^5\\
&+840ah_2^5c_1^2c_2^3c_3^2 + 630ah_2h_1^4c_2^2c_3^5 - 1080ah_2h_1^5c_2^2c_3^4\\
&+12600ah_2^5h_1^2c_1c_2c_3^3 - 360ah_2^5c_1^3h_1c_3^3 - 360ah_2^5c_2^4h_1c_3^2\\
&+280ah_2^5c_1^3c_2c_3^3 + 1260ah_2^5c_1^2c_2^2c_3^3 - 420ah_2^2c_1^3h_1^2c_3^5\\
&+3500ah_2^2h_1^3c_2^4c_3^3 - 50400ah_2^2h_1^2c_1^2c_2^3c_3^3 - 2250ah_2^2h_1^4c_2^4c_3^2\\
&-2100ah_2^2h_1^2c_2^4c_3^4 - 3150ah_2^2c_1^2c_2^4c_3^4 + 1575ah_2^2h_1^5c_1c_3^4\\
&-108ah_2h_1^5c_1^2c_2^4 + 84ah_2h_1^6c_1^2c_2^3 + 840ah_2^5c_1c_2^3c_3^3\\
&-4320ah_2^5c_1c_2^3h_1c_3^2 - 2160ah_2^5c_1c_2h_1c_3^4 - 896ah_2^5c_1^3h_1^3c_2\\
&-4032ah_2^5h_1^3c_1^2c_2^2 + 4200ah_2^2h_1^5c_2^3c_3^2 - 16800ah_2^2c_1^3c_2^3h_1^2c_3^2\\
&+3150ah_2^2h_1^5c_1^2c_3^3 + 210ah_2^5c_1c_2^4c_3^2 + 7ah_2h_1^6c_3^5 + 2310ah_2^5h_1^6c_3\\
&+7ah_2^5c_1^3c_2^4 + 3780ah_2^5h_1^4c_3^3 - 3024ah_2^5h_1^5c_2^2 + 11340ah_2^5h_1^4c_2^2c_3\\
&+105ah_2^5c_1^2c_2^4c_3 - 24000ah_2^3h_1^3c_1^2c_2^3c_3 + 42000ah_2^3h_1^2c_1^2c_2^3c_3^2\\
&+5250ah_2^3h_1^2c_1^2c_2^4c_3 - 6720ah_2^3h_1c_1^2c_2^4c_3^2 - 12000ah_2^3c_1^3h_1^3c_2^2c_3\\
&+21000ah_2^3c_1^3h_1^2c_2^2c_3^2 + 7000ah_2^3c_1^3h_1^2c_2^3c_3 - 8960ah_2^3c_1^3h_1c_2^3c_3^2\\
&-4032ah_2^3c_1^2c_2^2h_1^5 + 1512ah_2^3c_1^2c_2^2c_3^5 - 4032ah_2^3h_1c_1c_2^2c_3^5\\
&+42000ah_2^3h_1^4c_1^2c_2c_3^2 + 21000ah_2^3h_1^2c_1^2c_2c_3^4 - 48000ah_2^3h_1^3c_1^2c_2c_3^3\\
&-13440ah_2^3h_1^5c_1^2c_2c_3 - 26880ah_2^3h_1^5c_1c_2c_3^2 - 24000ah_2^3h_1^3c_1c_2c_3^4\\
&+42000ah_2^3h_1^4c_1c_2c_3^3 + 5040ah_2^3h_1^6c_1c_2c_3 - 6720ah_2^3h_1^5c_1c_3^3\\
&+2520ah_2^3h_1^6c_1c_3^2 + 5250ah_2^3h_1^4c_1c_3^4 + 1050ah_2^3h_1^4c_1c_2^4\\
&+4200ah_2^3h_1^4c_1^2c_2^3 - 1200ah_2^3h_1^3c_1^2c_2^4 + 2100ah_2^3c_1^3h_1^4c_2^2\\
&-1600ah_2^3c_1^3h_1^3c_2^3 - 72000ah_2^3c_1^2c_2^2h_1^3c_3^2 - 20160ah_2^3c_1^2c_2^2h_1c_3^4\\
&+63000ah_2^3c_1^2c_2^2h_1^2c_3^3 + 31500ah_2^3c_1^2c_2^2h_1^4c_3 + 63000ah_2^3h_1^4c_1c_2^2c_3^2\\
&+31500ah_2^3h_1^2c_1c_2^2c_3^4 - 72000ah_2^3h_1^3c_1c_2^2c_3^3 - 20160ah_2^3h_1^5c_1c_2^2c_3\\
&-224ah_2^3c_1^3h_1c_3^5 - 896ah_2^3c_1^3c_2h_1^5 + 336ah_2^3c_1^3c_2c_3^5 + 350ah_2^3h_1^4c_3^5\\
&+336ah_2^3h_1^6c_2^3 + 35ah_2^5c_1^3c_3^4 + 1512ah_2^5h_1^4c_2^3 + 105ah_2^5h_1^2c_3^5
\end{aligned}
$$

$$-224ah_2^3c_2^4h_1^5 + 7ah_2^6h_1c_3^5 - 140ah_2^6h_1^2c_3^4 - 3ah_2^6c_1c_3^5$$
$$+42ah_2^5c_2^2c_3^5 - 3ah_2^6c_1^2c_2^4 - 4ah_2^6c_1^3c_2^3 - 2100ah_2^6h_1^4c_3^2$$
$$+1386ah_2^6h_1^5c_1 - 1260ah_2^4h_1^6c_2^2 - 630ah_2^4h_1^6c_1^2 - 140ah_2^4c_2^4c_3^4$$
$$+350ah_2^4h_1^3c_3^5 - 896ah_2^3h_1c_2^3c_3^5 + 3360ah_2^3h_1^6c_2c_3^2 + 7000ah_2^3h_1^4c_2c_3^4$$
$$-8960ah_2^3h_1^5c_2c_3^3 - 13440ah_2^3h_1^5c_2^2c_3^2 - 12000ah_2^3h_1^3c_2^2c_3^4$$
$$+21000ah_2^3h_1^4c_2^2c_3^3 + 2520ah_2^3h_1^6c_2^2c_3 - 6720ah_2^3h_1^5c_1^2c_3^2$$
$$-6000ah_2^3h_1^3c_1^2c_3^4 + 10500ah_2^3h_1^4c_1^2c_3^3 + 1260ah_2^3h_1^6c_1^2c_3$$
$$-4480ah_2^3c_1^3c_2h_1c_3^4 + 14000ah_2^3c_1^3c_2h_1^2c_3^3 + 7000ah_2^3c_1^3c_2h_1^4c_3$$
$$+12600ah_2h_1^4c_1^2c_2^3c_3^2 + 31500ah_2^2h_1^3c_1c_2^2c_3^4 + 42ah_2^6c_1^3h_1c_2^2$$
$$-840ah_2^6h_1^2c_1^2c_3^2 + 1008ah_2^6h_1^3c_1^2c_2 + 2100ah_2^4c_2^3h_1c_3^4$$
$$-27000ah_2^4h_1^2c_1c_2^3c_3^2 + 12600ah_2^4h_1c_1c_2^3c_3^3 - 1120ah_2^3c_1^3c_2^4h_1c_3$$
$$-26880ah_2^3h_1c_1^2c_2^3c_3^3 + 2520ah_2^3c_1^3c_2^2c_3^4 - 48000ah_2^3c_1c_2^3h_1^3c_3^2$$
$$-13440ah_2^3c_1c_2^3h_1c_3^4 + 42000ah_2^3c_1c_2^3h_1^2c_3^3 + 21000ah_2^3c_1c_2^3h_1^4c_3$$
$$+350ah_2^3c_1^3c_2^4h_1^2 + 840ah_2^3c_1^3c_2^4c_3^2 + 1260ah_2^3c_1c_2^4c_3^4$$
$$+5040ah_2^3c_1^2c_2^3c_3^4 + 3360ah_2^3c_1^3c_2^3c_3^3 + 2520ah_2^3c_1^2c_2^4c_3^3$$
$$-6720ah_2^3h_1c_1c_2^4c_3^3 - 13440ah_2^3c_1^3h_1c_2^2c_3^3 - 420ah_2^4h_1^4c_2^4$$
$$+3780ah_2^4h_1^5c_3^3 + 378ah_2^4c_1^3h_1^5 - 112ah_2^4c_2^3c_3^5 + 1512ah_2^4h_1^5c_2^3$$
$$-2100ah_2^4h_1^4c_3^4 - 28ah_2^4c_1^3c_3^5 - 6750ah_2^4c_1^3c_2^2h_1^2c_3$$
$$-2250ah_2^4c_1^3h_1^2c_3^3 - 1680ah_2^4c_1^3c_2^2c_3^3 - 12600ah_2^4h_1^4c_1^2c_3^2$$
$$+10500ah_2^4h_1^3c_1^2c_3^3 - 2100ah_2^4c_1^3h_1^4c_3 + 2100ah_2^4c_1^3c_2^2h_1^3$$
$$+5670ah_2^4h_1^5c_1^2c_3 + 6804ah_2^4h_1^5c_1c_2^2 - 2520ah_2^4h_1^6c_1c_2$$
$$+4536ah_2^4h_1^5c_1^2c_2 - 1680ah_2^4c_1^3h_1^4c_2 - 7560ah_2^4h_1^4c_1^2c_2^2$$
$$-6000ah_2^3h_1^3c_1c_2^4c_3 + 10500ah_2^3h_1^2c_1c_2^4c_3^2 - 1200ah_2^3h_1^3c_1c_3^5$$
$$+1008ah_2^3h_1^6c_1^2c_2 + 14000ah_2^3h_1^4c_2^3c_3^2 + 7000ah_2^3h_1^2c_2^3c_3^4$$
$$-16000ah_2^3h_1^3c_2^3c_3^3 - 4480ah_2^3h_1^5c_2^3c_3 - 4000ah_2^3c_2^4h_1^3c_3^2$$
$$-1120ah_2^3c_2^4h_1c_3^4 + 3500ah_2^3c_2^4h_1^2c_3^3 + 1750ah_2^3c_2^4h_1^4c_3$$
$$+3500ah_2^3c_1^3h_1^4c_3^2 + 1750ah_2^3c_1^3h_1^2c_3^4 - 4000ah_2^3c_1^3h_1^3c_3^3$$
$$-1120ah_2^3c_1^3h_1^5c_3 + 1512ah_2^3h_1^6c_1c_2^2 - 2688ah_2^3c_1c_2^3h_1^5$$
$$+1008ah_2^3c_1c_2^3c_3^5 - 16000ah_2^3c_1^3c_2h_1^3c_3^2 + 22680ah_2^4h_1^5c_1c_2c_3$$
$$-1260ah_2^2c_1^3c_2^2c_3^5 + 1050ah_2^2h_1^3c_1^2c_3^5 + 525ah_2^2h_1^5c_2^4c_3$$
$$-900ah_2^2h_1^4c_2c_3^5 + 2100ah_2^2h_1^3c_2^2c_3^5 - 1680ah_2^2h_1^2c_2^3c_3^5$$
$$+11340ah_2^2c_1^2c_2^4h_1c_3^3 + 5250ah_2^2c_1^2c_2^4h_1^3c_3 - 4200ah_2^2c_1^3c_2^3c_3^4$$
$$-840ah_2^2h_1^6c_1c_3^3 - 7560ah_2^5h_1^5c_1c_3 - 4480ah_2^5h_1^3c_2^3c_3$$
$$-25200ah_2^4h_1^4c_1^2c_2c_3 + 7000ah_2^4c_1^3h_1^3c_2c_3 + 31500ah_2^4h_1^3c_1^2c_2^2c_3$$
$$+21000ah_2^4h_1^3c_1c_2^3c_3 - 840ah_2^4c_1c_2^4c_3^3 + 11340ah_2^4h_1^5c_1c_3^2$$